D0629805
WITHDRAWN
No longer the property of the
Boston Public Library.
Sale of this material benefits the Library.

Places for Dead Bodies

Places for Dead Bodies

by Gary Hausladen

University of Texas Press Austin

PRINTED IN THE UNITED STATES OF AMERICA
FIRST EDITION, 2000

⊗ THE PAPER USED IN THIS BOOK MEETS THE MINIMUM
REQUIREMENTS OF ANSI/NISO Z39.48-1992 (R1997)
(PERMANENCE OF PAPER).

LIBRARY OF CONGRESS CATALOGING-IN-PUBLICATION DATA
HAUSLADEN, GARY, 1946–
PLACES FOR DEAD BODIES / BY GARY HAUSLADEN. —
1ST ED.
P. CM.
INCLUDES BIBLIOGRAPHICAL REFERENCES (P.) AND
INDEX.
ISBN 0-292-73127-2 (CLOTH : ALK. PAPER) — ISBN
0-292-73130-2 (PBK. : ALK. PAPER)
1. DETECTIVE AND MYSTERY STORIES, AMERICAN—
HISTORY AND CRITICISM. 2. POLICE IN LITERATURE.
3. DETECTIVE AND MYSTERY STORIES, ENGLISH—HISTORY
AND CRITICISM. 4. PLACE (PHILOSOPHY) IN LITERATURE.
5. GEOGRAPHY IN LITERATURE. 6. SETTING (LITERATURE)
I. TITLE.
PS374.P57 H38 2000
813'.08720932—DC21 99-047170

For Marilyn Elise, Theodore, Bradley, and Christina

Contents

Preface

The first inkling of an idea for this kind of study occurred to me while reading *Gorky Park* on a flight returning from Russia in 1982. I was fascinated by Martin Cruz Smith's ability to convey so vividly and accurately a sense of Moscow and the Soviet society I thought I knew so well, after having lived there for half a year with my wife and children. Some years later I would put that fascination into play with a meeting presentation, and an article on Moscow through the eyes of escapist literature was born. Who could resist taking the subject a step further? And one step led to another—to a more general treatment of place and police procedurals, and then finally to this book.

Truly this has been a labor of love. I have more than once felt guilty that so much enjoyment was to be had in what is supposed to be a painful and laborious task. Murder mysteries have been a favorite of mine as long as I can remember. As a child my favorite stories and movies included Sherlock Holmes, any film noir movie of the 1950s, and most any TV program that included a murder and its resolution as the primary plot.

I came to appreciate the power of murder mysteries to enrapture audiences. The sense of place as a geographical concept offers the opportunity to examine murder mysteries from a different angle and to deal with an increasingly popular academic subject—escapist literature as a measure of contemporary society, from a specifically geographical perspective. That I was a murder mystery fan long before I became a geographer probably shows in my treatment of novels in this book. I cheerfully and willfully plead guilty.

This book is written more for mystery fans than academics. But as I have discovered, these two are not necessarily exclusive. I am regularly startled by colleagues who admit to being ardent fans. They're often gracious even when I have to admit to excluding one of their favorite authors. So here we have it, I hope—a serious and important topic written as much for enjoyment and pleasure as for critiquing the ability of the police procedural genre to convey a sense of place. There is no guilt in indulging in this escapade into escapism. I only hope you enjoy reading this as much as I have enjoyed writing it.

Acknowledgments

Debts accrue to numerous people who have offered advice and encouragement. The production of a book-length manuscript from this material was an idea that dawned on me during my first visit to a mystery bookstore—Mystery Books of Washington, D.C. And so Jerry, since deceased, Barbara, Tina, and all the folks at Mystery Books, past and present, are owed thanks for initial and continued encouragement. In a field in which books go out of print and are difficult to find, I am always amazed at what is available in the many bookstores of Berkeley, California, especially Black Oak Books, Cody's, and Pegasus bookstores. For the numerous readers and listeners, student and nonstudent, who never waivered in enthusiasm, positive criticism, and new ideas, I confer thanks, especially to Paul Starrs; Bill Wyckoff; my wife, Marilyn; and my father-in-law, Ben McGuire, each of whom spent hours reading and commenting on this manuscript. I am indebted to Barbara Friedsam for her imagination and creativity in the production of the maps that precede the chapters. As well, three anonymous reviewers offered positive and useful comments and suggestions that enhanced the quality of the text. Where I have taken my own course in spite of their advice, I accept full responsibility.

Two supporters deserve special recognition—Dan Arreola, who introduced me to Shannon Davies of the University of Texas Press, and Shannon Davies herself, whose unswerving faith in this project ensured that the book would become a reality. I offer a debt of thanks as well to Sheri Englund, Leslie Tingle, Nancy Crumpton, and all the folks at U.T. Press.

This work first saw the light of day in two publications, which have given permission to use material previously published. "Murder in Moscow" was my first foray into publication of these materials and serves as the basis for Chapter 6. The original work was published in the *Geographical Review* 85 (January 1995): 63–78. A more generic treatment of the subject appeared in the *Journal of Cultural Geography* as "Where the Bodies Lie: sense of place and police procedurals," vol. 16 (Fall/Winter 1996): 45–64. At the very least, it serves as the basis for Chapters 1 and 2; in fact, it presents the rationale for the entire book. The police procedural series used as examples in these articles have found their way into the appropriate chapters. Thanks to both the American Geographical Society and the Popular Press for these permissions.

Introduction

Chapter 1

The contemporary mystery scene is a stimulating place for reader and writer alike.
—JON L. BREEN, IN The Fine Art of Murder[1]

This book examines how police procedurals are used to clearly and effectively convey a fundamental geographic and literary theme—"sense of place." Not only is sense of place essential to creating an authentic locale for the plot of the novel, an authenticity that is absolutely necessary to preserve credibility, it also serves as a source, sometimes the sole source, for exposing thousands of readers to other places. Popular, escapist literature in particular is sated with insidious but powerful insight into places and cultures, some exotic, some familiar. Simply because literature may be popular and escapist does not diminish its impact. People may read popular literature to escape, but they cannot escape from the descriptions and impressions of places that are such an integral part of the plot. In these cases, fiction and geography don't just meet; they unite in imaginative and provocative ways to further the agendas of each.

Most of this book focuses on how geography, and particularly sense of place, is used to further the goals of literature to produce an effective and enthralling plot. Concomitantly, fiction, specifically the police procedural genre, is used as a powerful communicator of sense of place. The question that arises concerns why fiction is such an effective vicarious conveyer of sense of place.

> [A] person often brings to literature an attitude that is more relaxed, more responsive, less inclined to prejudgment than he or she might bring to a textbook. Fiction . . . does encourage the mind to explore more willingly and freely.[2]

If true for fiction in general, this is even more true for mysteries, when the need for entertainment often challenges the reader to read insatiably to deduce the solution of the mystery before it is revealed by the author. Seldom sedate, but always escapist, the police procedural, as with all detective fiction, requires the reader to think, to match wits with those of the investigator in the narrative, and to participate in the investigation. It has evolved as a particularly popular genre of the murder mystery, a genre in which the demands of the police procedural format and the evocative power of place engage each other in a cohesive and mutually beneficial relationship that makes many of these novels PLACE-BASED POLICE PROCEDURALS.[3]

The sense of place found in these police procedurals is more than simply backdrop for the plot, the use of place more than simply a setting for the crime. Rather, place is an essential ingredient in the commission, discovery, and resolution of the crime. The use of place in mystery novels is just one specific example of how place is employed in all narrative fiction to further the effectiveness of the plot. In these kinds of mysteries, much of the intrigue is a function of locale. Place becomes an essential—maybe *the* essential—plot element. Nowhere else could these kinds of murders have occurred; they are culturally and contextually specific. Without a sense of place, be it the Navajo country of Tony Hillerman, the Yorkshire of Peter Robinson, the Australian Bush of Arthur Upfield, or the Moscow of Martin Cruz Smith, the plot of these mysteries is needlessly enigmatic. Scores of police procedurals can be classified as place-based.

Herein lies the purpose of this book—an examination of how and how effectively authors weave place into the tapestry of the plot of police procedurals, and how they elicit in readers a sense of place through popular, escapist fiction.

Part of the allure of the place-based police procedural is provided by a tension that necessarily arises in writing about places between what is familiar and what is not. The reader is exposed to images, sounds, smells, behaviors, and cultures that are new and unfamiliar, all the while being fed situations and circumstances that are recognizable and familiar, providing the basis for empathy and personal association. Even in exotic places, the key to success is in a careful and realistic depiction of the human condition, making our heroes and heroines real people with foibles and shortcomings like the rest of us, and who struggle with many of the same problems we all encounter in our daily lives. The trick is to engage the imagination of the reader. It is in the reading process itself, in the intense interaction between reader and text "that the special quality of the tale of the detective becomes evident."[4] The more the plot can captivate with what is familiar, the more easily it can entreat a reader to accept the unfamiliar. Engaging the imagination of the reader is essential to solidifying the bond between reader and protagonist. Once the bond is forged, the author has clear sailing to feed us sense of place.

This book examines the sense of place found in place-based police procedurals from two complementary perspectives, namely the perspectives of author and reader. How an author uses place as an integral part of the plot has everything to do with enhancing the effectiveness of a story. At the same time, place-based police procedurals create a complex and realistic sense of place in a reader's mind. Two basic criteria govern the selection of mysteries to be discussed. First, they must be police procedurals, with some generous tolerances. Second, the series selected must be placed-based—place must be an important plot element.

Most police procedurals turn into series. The series examined in this study range in number from two (P. M. Carlson's Marty Hopkins series) to twenty-nine (Arthur Upfield's Napoleon "Bony" Bonaparte series). The fact that most authors have produced series based on the same police officer is an extension of the effectiveness of the police procedural. Once readers identify with a given protagonist or group of protagonists, they became committed and dedicated fans of the entire series, often waiting impatiently for the next novel to find out how their heroes and heroines have fared. In *Finding Moon,* which breaks away from his Navajo tribal police series, Tony Hillerman felt compelled to soften the disappointment experienced by avid Jim Chee/Joe Leaphorn fans who had been waiting patiently to find out how our heroes' lives were progressing, and to assuage any fears that the series may be at an end. Hillerman is well aware of the source of his literary success:

> To my fellow desert rats, my apologies for wandering away from our beloved Navajo canyon country. The next book will bring Jim Chee and Joe Leaphorn of the Tribal Police back into action.[5]

As promised, in late 1996 Hillerman returned to the Navajo surrounds to continue the exploits of Leaphorn and Chee in *The Fallen Man.* His "fellow desert rats" did not fail him, providing *The Fallen Man* with a place at the top of the *New York Times* best-seller list for several weeks.

William Marshall learned his lesson the hard way. In the mid-1980s, he departed the ever-popular streets of Hong Bay, Hong Kong, and the exploits of Harry Feiffer and officers of the Yellowthread Street police station, to take up fictional residence in the Philippines. After fourteen years, Marshall had tired of the Feiffer series, believing that he had taken it as far as it could be developed. Fans, however, believed differently and sent Marshall a clear message by not buying the Manila Bay series. Marshall got the message and returned posthaste to Yellowthread Street with *Frogmouth* in 1987. What is interesting in the Marshall case is that not only had the public become addicted to the Feiffer series, so also had Marshall himself, apparently. The Manila Bay series lacks the humor and sarcasm that characterizes the antics of the Yellowthread Street gang.[6] Marshall continues to work on improving the Manila Bay series; yet, to the relief of a dedicated and expectant following, the Yellowthread Street series continues on into the late-1990s.

This addiction speaks to the value of developing a series based on the same characters and places. Over time, the author is able to add, reinforce, and change earlier images and impressions. The picture and setting become more complex, more real, and more familiar as the series evolves. The same places are revisited, new places are visited, and the sense of place is strengthened and expanded. A police procedural series is only one way to approach sense of place, but it is clearly effective.

The need to examine only place-based series and to separate the sense of place from character and plot development necessarily excludes some of the most commonly used venues for murder and some very popular series. Ed McBain, Hillary Waugh, and K. C. Constantine exemplify some of the better-known masters of the art whose works, based in New York City (the fictional Isola), Stockford, Connecticut, and Rocksburg, Pennsylvania, respectively, are not included in this book. European favorites like George Simenon and Donna Leon are also excluded. This exclusion in no way suggests that their series are not some of the finest or most popular police procedurals. They are. It means, however, that they fail to meet the place-based criterion of this study. For these giants of the police procedural, generic settings are used as backdrops, and the unique characteristics of places do not figure prominently in the commission, discovery, or resolution of their crimes.

For the most part, readers have been inundated with images of New York, Chicago, San Francisco, and Los Angeles, where most TV series, movies, and novels are set. These have become generic places, where the place itself is not essential to the development of the plot. This parallels certain successful TV police series like *Hill Street Blues,* in which it is never revealed explicitly where the station is located. Most viewers assumed that *Hill Street Blues* took place in New York City. In fact, it could have been any large northeastern U.S. metropolis. Identifiable features of New York City that could have been used to further plot development were not. Compelling character development, a team of police officers working on a number of different cases, employing real-life police procedures, categorize this series as a police procedural, and an effective one at that. Yet, clearly it was not place-based. Interestingly enough, *Dragnet,* the "father" of radio and TV police procedurals and often credited with initiating the demand for the development of the entire genre, was clearly based in Los Angeles. There was no ambiguity of place, even if the environs of LA were not employed as extensively as they might have been.

New York, Chicago, San Francisco, and Los Angeles are familiar places that provide the benchmark for readers evaluating other places, which I might take great liberty to refer to as "exotic places" for murder—places that the vast majority of the reading public are not familiar with and, as a result, are more accepting of the authors' interpretations. The less familiar a person is with a place, the more he or she has to rely on another's perspective. This applies to simple descriptions of landscape, social convention, cultural artifacts—in other words, all human and physical characteristics of a place.

At least one London-based police procedural series presents just the opposite scenario. P. D. James is a giant of British murder mysteries; yet, recently her novels have come under attack for weak character development, which is an interesting criticism given the great extent to which she elaborately and painstakingly crafts each character. *Original Sin* won high praise, however, for its sense of place, where the role of a single place, Innocent House, is an extreme example of place becoming *the* essential plot element. P. D. James emerges, then, as the only real representative of the London-based police procedurals. Most other police procedurals set in London fall into the generic places category.

A similar schism is found in Italy, the locale for a number of mystery series. Comparing the Venice of Michael Dibdin and Donna Leon is to examine two excellent police procedural series. But when it comes to being place-based, the Dibdin novels excel in capturing the essence of place-based police procedurals; whether the Vatican, Rome, Venice, or Naples, location is absolutely necessary for the respective plots of each. Although Leon includes a number of excellent descriptions of Venice, these descriptions are not necessary plot ingredients. The police procedures of Dibdin's Aurelio Zen, on the other hand, are depicted in far more detail and far more realistically than those of Leon's Guido Brunetti. The Zen novels are included in this book, and the Brunetti series excluded, even though I highly recommend Donna Leon to any fan of the murder mystery. Her novels simply do not meet the criterion of being place-based.

It should be noted that certain series have a historical and personal advantage—the first introduction to a place, done well, tends to remain a reader's favorite. Case in point is the New Orleans of James Lee Burke. I was mesmerized by his flavor of the "Big Easy," so much so that although his series broke away from the police procedural genre after the first novel, I had decided to use the exploits of Dave Robicheaux as representative of sense of place in New Orleans in any case. Robicheaux' return to police work after the second novel relieved my need to loosen the rules. The same can be said of Paco Taibo, since only one of his novels in translation is a true police procedural. But, as the sole representative from Latin America, and as the purveyor of a very notable sense of place, allowances are made.

Series suggested by other avid murder mystery fans carry a similar advantage. In addition to Burke, Upfield, Marshall, Dibdin, Lindsey, Gur, Dexter, and Melville were all authors recommended by people with contagious enthusiasm, whose opinions I value. An author might like to think these would have been included in any event; but being suggested by friends and colleagues as being particularly strong in their sense of place had its effect.

The result here is a selective sampling of some of the best place-based police procedurals from a representative number of places—thirty-three series in all. This is not an exhaustive study, although that was an original, naive, and certainly overly ambitious intent. More appropriately, the purpose now is to discuss a new and fascinating approach to murder mysteries, while explicitly identifying the power of popular literature in conveying to readers a sense of place.

Chapter 2 sets the context for the study and introduces a specifically place-based approach to police procedurals, focusing on their ability to generate a sense of place. Description, dialog, iconography, and attention to detail emerge as valuable literary devices to integrate sense of place into the plot.

With the spadework done, we launch into the police procedurals themselves in Chapters 3–9. They are divided geographically. Chapter 3 begins the journey in North America, examining the Navajo country of Tony Hillerman, Jean Hager's Cherokee country, the Houston and Latin America of David Lindsey, James Lee Burke and Julie Smith's New Orleans, the Lawrence County, Indiana, of P. M. Carlson, the Seattle of J. A. Jance, Susan Dunlap's Berkeley, and the Canadian north of Scott

Young. The tour of the Americas ends in Mexico with the translated works of Paco Taibo II.

Chapter 4 returns to the British Isles where the police procedural has of late gained stature equal to the traditional British whodunits. There is the London of P. D. James, the Oxford of Colin Dexter, the Yorkshire of Peter Robinson, the Glasgow of Peter Turnbull, and Bartholomew Gill's Dublin.

Chapter 5 moves across the Channel to the European continent where we explore deadly artistry set in Michael Dibdin's Italy, Nicolas Freeling's provincial France, Janwillem van de Wetering's Amsterdam, and Sjowall and Wahloo's Stockholm.

The next four chapters offer some untraditional contexts for murder mysteries. Russia is the scene of the murders examined in Chapter 6, where the Moscow of Martin Cruz Smith and Stuart Kaminsky provides places for dead bodies. In Chapter 7, we visit the Orient where Seicho Matsumoto and James Melville offer two quite different approaches to murder in Japan, William Marshall guides through the streets of Hong Kong, and Christopher West offers a relatively new series set in communist Beijing. In Chapter 8, we travel to other places for murders in Batya Gur's Israel, H. R. F. Keating's Bombay, James McClure's South Africa, and the Australia of Arthur Upfield. An interesting and intriguing twist to the murder mystery is uncovered in Chapter 9, which examines murder mysteries based in the historical past. These include the Imperial Rome of Lindsey Davis, the seventh-century China of Robert Van Gulik, the Victorian England of Anne Perry, and Michael Pearce's turn-of-the-century Cairo.

Chapter 10 concludes the investigation with some final comments about future places for uncovering mystery, murder, and mayhem, and about the "authenticity" of the senses of place presented in the murder mystery novels. Do they, as a whole, "get it right?" How do we assess the accuracy of the places portrayed? Is our predicament any different from similar predicaments in other more serious kinds of literature? The Fictional Works Cited at the end of this book provides complete lists of the novels of the series discussed in this book. Because of continuing additions to many of these series, I can only guarantee that the lists are complete as of summer 1999.

No doubt, numerous queries will ask why I failed to include "what's his name, my favorite detective." I am well aware that for each one written about, there are two or three others I haven't addressed. It is impossible to include all series in this one volume. The sampling of police procedurals could easily have been two or three times as large. Keep in mind that a number of excellent series have been excluded because they failed to meet the basic criteria or because the place they depict is already well represented. My primary purpose is to bring to light a different, characteristically geographical approach to the police procedural and to underscore the power of the murder mystery novel to convey a sense of place. The series used to help make that point provide a sampling, if an extensive one, of what is available. Further inquiries into other places for murders will be the grist for future investigations and many additional hours of pleasurable reading.

The Evolution of the Place-Based Police Procedural

Chapter 2

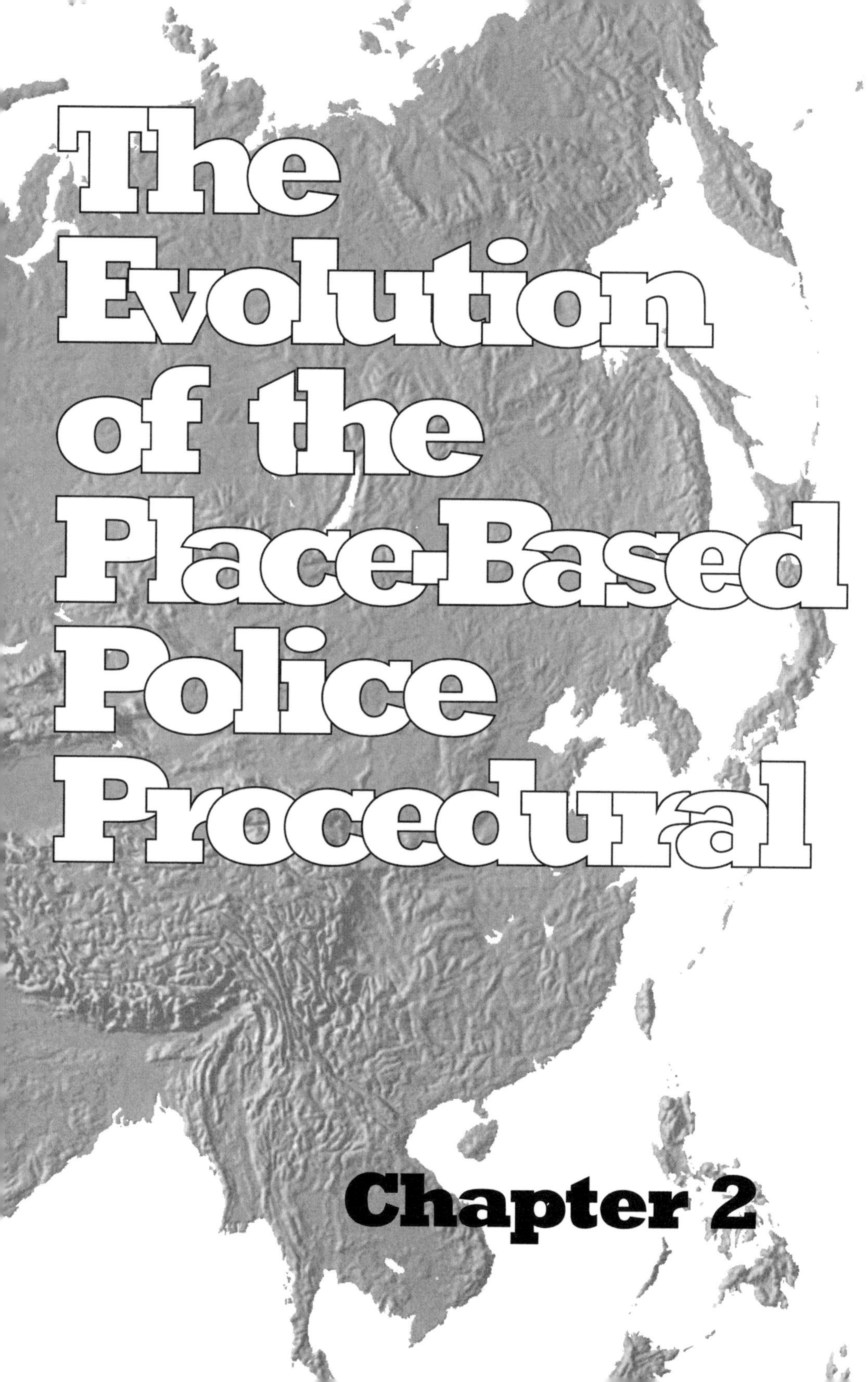

Readers of detective fiction have always craved authentic details of police work, and authors have long endeavored (or at least pretended) to provide them . . .
—JON L. BREEN, IN The Fine Art of Murder[1]

Edgar Allan Poe's "Murders in the Rue Morgue," published in 1841, is widely accepted as the first modern detective story and Poe's detective Auguste Dupin as a prototype of the super sleuth who would be perfected four decades later by Arthur Conan Doyle's Sherlock Holmes. Even Charles Dickens, who created "Inspector Bucket of the Detective" in *Bleak House* in 1852, dabbled in the detective genre; for all that, Dickens was careful to market it as a mainstream serial, not as a detective story. Respectability, and sales, were still questionable for the fledgling mystery genre.

Without question, however, it was Arthur Conan Doyle who popularized the art form invented by Edgar Allan Poe. In 1887 *A Study in Scarlet* introduced the world to Sherlock Holmes, who would become the quintessential master detective of the murder mystery and the source of the classic murder tale. His popularity seems to endure unabated. In addition to popularizing the murder mystery, Holmes provided a prototype for the genre that evolved during the classical age of murder mysteries in the 1920s and 1930s. Tracing the origins of detective fiction back to Poe and Conan Doyle, of course, ignores that the Chinese were producing detective fiction as early as the seventeenth century. Robert Van Gulik received his inspiration for writing the Judge Dee series after uncovering seventeenth-century murder mysteries during his research on Imperial Chinese literati.

Some have even argued that the very first murder mystery was *Oedipus Rex,* singled out by Aristotle in *The Poetics* as the founding work of Western literary theory. If true, it appears that the ties between detective fiction and "serious" literature go back to the very origins of narrative fiction.

> [T]his particular plot [*Oedipus Rex*] derives its formal brilliance from a concealed secret, in fact the identity of a murderer, which

> is raised as the central problem at the beginning and solved at the end, with the interim being totally taken up by the inquiry.[2]

Could it be that the first murder mystery was written in 430 B.C.? Possibly detective fiction is not such a recent genre as we have long assumed. One of the intriguing aspects of *Oedipus Rex* is that Sophocles accomplished a truly amazing feat by creating a character, Oedipus, who was the criminal, the detective, and, with a stretch of the imagination, the victim all rolled into one! There are numerous examples of one character combining two of these roles, e.g., detective and victim or detective and criminal, but seldom if ever all three. *Oedipus Rex* is also the first example of " 'uroboric' fiction, one which uncovers the events leading up to its own beginning," which is the basic format for all detective fiction.[3]

The truth about its historical roots notwithstanding, detective fiction as we know it has a decidedly Western flavor, especially tied to the English-speaking world. The classic amateur detective mystery, which traces its "golden age" to the period between the two world wars, included

> the eccentric and omniscient detective, the story narrated by his less astute comrade, the police who see everything but observe little, the staged ruse to force the perpetrator's hand, the clues placed so that the reader should be able to follow the detective's reasoning (but he doesn't), the surprise solution, the explanation, at some length, in which the reader is shown how simple it was to determine the solution.[4]

The rules were formulaic, the solutions often mechanical, and the resolutions ritualistic. But detective mysteries sold by the gross. Among the many fine authors who contributed to the age of the classic British whodunit are superstars like Agatha Christie, Dorothy Sayers, and G. K. Chesterton, to name but three. From their fictional pages emerged such unforgettable characters as Miss Marple, Hercule Poirot, Lord Peter Wimsey, and Father Brown. Many of these characters remain popular to this day on television and the silver screen.

This basic format was transported to the United States, where it found a warm reception in the east. The American variant was promulgated by Ellery Queen (Frederic Dannay and Manfred Lee), S. S. Van Dine, and Rex Stout. And amateur detectives such as Ellery Queen, Philo

Vance, and the estimable Nero Wolfe became both well-known literary characters and household names.

Yet, critics and readers alike opined that the format for detective novels had grown so rigid and unrealistic that they proved standardized exercises in puzzle solving rather than novels in any literary sense. They lacked the humane values of the classical novel. This changed in the 1920s and 1930s in the western United States with the emergence of two particularly gifted crime writers—Dashiell Hammett and Raymond Chandler. Their more realistic approach to crime writing paralleled the emergence of the paperback book and inexpensive mass circulation. These two innovations changed the face of mystery writing forever, exposing ever greater numbers of fans to the genre. If there were still issues about whether murder mysteries were stylistically serious literature, there was no question about their increasing impact on American society. The renowned sociologist Fredric Jameson, after admitting with some sense of apology that he was an ardent fan of murder mysteries, described Raymond Chandler "as a painter of American life . . ." who presented his paintings "in fragmentary pictures of setting and place, fragmentary perceptions which are by some formal paradox somehow inaccessible to serious literature."[5] For Jameson and others, murder mysteries influenced the public and provided for serious topics in unthreatening prose in ways that serious literature could not. The question of whether murder mysteries were "serious" literature was moot.

The new, western American style of detective novels, the "hard-boiled school," differed from its classic forebears. First, it was more realistic. Murder was not conducted by set rules, pitting highly sophisticated murderers against super sleuth, if often amateur, detectives. The murders were often grisly, copious in number (as in Hammett's *Red Dawn*), the murderers bungling, and the detectives, by now professionals, were human. Agatha Christie, Ellery Queen, and Rex Stout made the transition with mixed success. They were joined by such popular writers as Ross MacDonald, John D. MacDonald, and later by Mickey Spillane, whose novels took a 1950s turn with an emphasis on sex, sadism, and sensationalism. Other genres developed as well—spy novels, serial killer mysteries, suspense novels. The detective novel was one of a number of variations of the mystery novel. Yet in many ways these novels still tested the reader's ability to suspend disbelief—ignoring real life. Like their classical amateur predecessors, the professional private eyes of the hard-boiled school were virtually free of legal restraint . . . were laws unto themselves . . . drank heavily while suffering not the slightest impairment of their facilities . . . and encountered a female population consisting almost totally of beautiful blond nymphomaniacs.[6]

The Police Procedural Genre

In the 1940s and 1950s, the police procedural evolved as the most recent adaptation to detective fiction. It was, for the most part, a reaction to the rigidity of the classic murder mysteries. The police procedural brought more realism to the mystery yarn. Unlike the heroes of Doyle, Christie, Chandler, Hammett, or Spillane, the police detective was not superhuman. He had to abide by rules and regulations, not all clients were insatiable bombshells, and invariably there was life outside the job. The police procedural genre has been compared to the other detective fiction variants that preceded it:

> Where the classic detective solves mysteries through the use of his powers of observation and logical analysis, and the private detective through his energy and his tough tenacity, the detective in the procedural story does those things ordinarily expected of policemen, like using informants, tailing suspects, and availing himself of the resources of the police laboratory.[7]

This variant of detective fiction produced dozens of popular writers. The initial American police procedural is widely attributed as Lawrence Treat's *V as in Victim,* published in 1945, which was a singular effort. The trend toward police procedurals as a popular form of the murder mystery was initiated in 1952 with Hillary Waugh's *Last Seen Wearing.* In Britain, J. J. Marric's *Gideon's Day* (1955) is credited as the first British police procedural, although some bestow the honor on Maurice Procter's *The Chief Inspector's Statement* (1952).

The popularity of the police procedural as a literary genre was provoked by the popularity of *Dragnet,* which was a popular radio series from 1949 to 1951. It became the first police procedural television series in 1952, where it would remain a mainstay of prime time until 1959. The series was revived in 1967 to entertain viewers for another three years.

Dragnet still serves as an excellent example of the basic components of the procedural approach to crime detection. The total impact of *Dragnet* must include not only the millions of police procedurals sold over the past forty-five years, but also the promulgation of numerous other police procedural TV series, including such successful ones as *Adam-12, Cagney and Lacey, CHiPs, Hawaii Five-O, Hunter, In the Heat of the Night, Miami Vice, Starsky and Hutch, The Streets of San Francisco,* and the long rerunning *Columbo.* More recently, a new wave of police procedurals have rekindled America's fascination with real-life police procedures, including *Hill Street Blues, Homicide, Nash Bridges, NYPD Blue,* and *Law & Order.* More sensationalized, less realistic, yet wildly popular offerings include *Pacific Blue, Silk Stalkings,* and *The Big Easy.* And the list goes on. The fact that a number of literary police procedurals have been turned into TV series or movies attests to the continuing symbiotic relationship between TV and popular literature, a relationship initiated by *Dragnet.*

The early police procedurals of Treat, Waugh, Marric, or Procter were not the first mysteries to employ a police detective as the main character; rather, they were the first mysteries to focus on the real-life processes of police detection. If the "Great Detective" tradition provided a gifted and brilliant individual who solved mysteries on his own and often was brought in from the outside, the procedural policeman was a plodder, albeit a cunning plodder, who worked as part of a team, depended on "cop sense" and technology, and generally was a member of the community where the crime was committed. The major change was the focus on actual procedures, whether of Scotland Yard, Bertillon's anthropometry, or the intricacies of the autopsy. The rest, as they say, is history. The police procedural has endured since the 1950s as a prolific and popular genre.

Four main elements distinguish the police procedural from other murder mysteries based on police protagonists. The most fundamental of these is attention to real-life police work and procedures. The procedurals deal not just with policemen, but with the "world of the policeman," involving all manner of rules and regulations.[8] This is the primary rationale for excluding such popular mystery series as Georges Simenon's Inspector Maigret. Simenon provides a policeman as hero, but Maigret most certainly does not operate in the world of the policeman. His is instead mainly of the world of a Sherlock Holmes, who more often solves cases by intuition rather than by police procedures.

For true procedurals, no matter how fantastic the story lines, one basic criterion is essential: "they must get the procedure right."[9] They also must get the geography and sense of place right as well. Because so much police work is ordinary, the key to a writer's success in this genre is turning the mundane into something exciting and suspenseful. The author "must create excitement out of the unexciting. He must make the mundane interesting. He must extract suspense from the routine."[10] This is the true art of the writer of the police procedural.

A second distinguishing element of these novels is the detection of crime based on teamwork. The police procedural is an ensemble performance. While there may be a lead detective, several colleagues are generally found in these novels, and often a core of investigators appears regularly throughout the series. These other policemen and women may not become as well known as the hero, but they do give opportunities of character development that, in turn, add to the realism and help readers identify with the investigators in the cases.

For many of these novels, the corps of colleagues allows the author to introduce another touch of realism and the third distinguishing element—several criminal investigations being conducted simultaneously. Oddly, this is a front energetically adopted in the 1990s television police shows but rarely in film. Quite clearly, the hero is conducting the main investigation; yet colleagues may be off investigating other crimes, often not other murders. Sometimes seemingly unrelated cases come together by the end of the novel as parts of a single crime. In other cases, they remain totally unrelated throughout—simply an extension of the intent to create a real-life environment.

The fourth distinguishing element also is meant to enhance the realism of the investigation—the description of the murder itself. The description of the body is often gruesome, distasteful, and inelegant. It can be simple, complex, professional, or bungled. Details of the murder provide a different kind of avenue to say something about the murderer, and also something about the hero and/or his colleagues. Descriptions of the murder, the murder scene, and the dead body are opportunities to combine outright description with individual reactions to further the development of character traits. There is something about the humanity of the characters, once again, that draws us to them as fellow human beings.

The success of the procedural format has spawned variations and imitations, including the attorney procedural, with the writings of Erle Stanley Gardner, A. E. Voelker, and Scott Turow; the coroner procedurals, as found in the writings of Patricia Cromwell; and one might even add writing can carry within it descriptions of places that are)est available: "I must insist that some of the best travel today occurs in the midst of spy and detective fiction."[12]

)r Winks the creation of places in detective fiction plays an .e. After stressing the necessity of accuracy in descriptions the demand for careful research to ensure this accuracy, he asic concept of this book—that place matters in numerous ls. In referring to writers of foreign-based detective fiction: se 'exotic' writers lends verisimilitude to the plot by care- to local detail, so that one senses the books emerging from pes."[13]

:ey point applicable to place-based murder mysteries is that e is not simply a crowd pleaser, nor is it used solely to ad- l agenda. It is crucial to plot development. Without place, e makes no sense. A crime would be agonizingly trivial. ıs' comments that Sara Paretsky's novels capture the ability ysteries to continue the use of realistic characters while add- an important component. In the Warshawski series, "[The gain a good deal from being soaked in the dangerous, but :s it heartening rather than depressing, area of South Chi- . is high praise from one of Britain's most highly regarded

stylistic blueprint is also found in numerous nonmystery aking "a case for popular novels as a major determinant of ace images," James Shortridge distinguishes forty-two au- n 1800 and 1950 whose novels were "place-defining," provid- statements of perceived regional values."[15] No one was more n establishing an image of the American West than Zane ugh Grey's formulaic plots have been decried by critics, all e his vivid descriptions of the vastness, wildness, and mys- Western landscape . . . these images remain in force today."[16] ight be useful here to pause for a moment to make sure that on exactly what I mean by "sense of place." For the geogra- of place" is part of the concept of place that includes the elements of "location," "locale," and "sense of place":

le, the settings in which social relations are constituted se can be informal or institutional); *location,* the geographi- rea encompassing the settings for social interaction . . . and *e of place,* the local "structure of feeling." Place, therefore,

the private detective procedural, as epitomized by the works of Joseph Wambaugh. Although the main character is a hard-boiled detective, the plot revolves around unraveling the crime through the use of realistic detective procedures. A strength of each of these authors is the use of more detailed character development, like in the police procedurals, than is found in the earlier varieties of murder mysteries.

Fellow human beings to whom we can relate and empathize—this is the goal, and character development is the means to this goal. Character development, although not unique to the police procedural, is one of the keys to the success of the genre, in part, because it provides a powerful medium for realism. The typical hero of the police procedural is one with whom the reader can readily relate. The reader can associate with him or her as a fellow human being and can empathize with his or her plight. This empathy might derive from the hero's personal situation. In novels we find family men, divorcees, alcoholics or recovering alcoholics, debtors, invalids, and people struggling with complex changes in their lives. Situations are introduced in various ways throughout the novel as asides and complements to the main plot.

Character development plays a more important plot role than in earlier hard-boiled detective yarns. Often the personal lives of the characters run parallel to the story line, adding primarily to the development of the character. It is not unusual for personal lives and professional cases to become intertwined. Aspects of a character's personal life sometimes become part of the murder mystery, as in the case of James Lee Burke's Dave Robicheaux, whose wife is murdered in *Heaven's Prisoners,* or Susan Dunlap's Jill Smith, whose ex-husband's girlfriend's disappearance initiates the plot of *As a Favor.*

Also related to character development is how the protagonist solves the murder and the numerous people and situations encountered along the way. Trials and tribulations, frustrations, successes, and failures are encountered in dealing with the mundane. The manner of the investigation, of course, is presented as an integral part of the main plot. We may have never been involved in the detection of a crime, but we certainly have experienced the frustration and anger of dealing with people and circumstances that have made our lives more difficult. Either way, the key is to be able to relate to and associate with what the protagonist is experiencing—basically, the reader needs to like the hero or heroine.

The police procedural acquired legitimacy as literature with the increasing character development in novels. If earlier detectives were more caricatures than characters, the development of personality in the police procedural became an important part of bonding reader and protagonist, something unattainable in earlier detective stories. This had the added benefit of permitting a character to be used in multiple novels, developing a sometimes fanatical following.

An author had better create criminals we can relate to as well. At best, they are real people, rather than master craftsmen or sinister evil geniuses. Their motives are generally basic human motives—sex, money, power. And the criminal likewise has a history and a life that is understandable, if not condonable.

Just as it is possible to identify a number of elements common to most police procedurals, there is also a great deal of variety. When the identity of the murderer is revealed is always an intriguing part of the author's approach. We can find out whodunit early on and then spend most of the novel seeing if our police procedures will bring the culprit or culprits to justice, or we can find out the culprit at the same time as the protagonist, having spent the entire novel trying to match wits with the police in determining who the guilty party or parties might be.

Another difference in approach relates to the length. Shorter novels tend to run about 180–220 pages, and a few are even shorter. Longer novels exceed 250 pages, with some exceeding 400. This disparity gets at the heart of the discussion of the mystery story versus the novel. Originally, murder mysteries were written primarily as short stories, distributed in mystery magazines, and some even serialized in the Dickensian tradition. This continues to the present in the still popular *Ellery Queen's Mystery Magazine,* begun in 1941, and the *Alfred Hitchcock Mystery Magazine,* first published in 1956. Many of today's best-known mystery writers began with short stories in these magazines, and many continue to write short stories for them after they've become famous.

With the evolution of the genre from short stories into novels, length grew as did the popularity of the genre. Many, such as the Yellowthread Street series, the Kaminsky series, or the Dunlap series, come out at right around 180 pages, a number that translates into approximately sixty thousand words. The size is a pulp fiction style, popular with publishers because the costs of production go up dramatically with larger books. The public appetite is changing also, however, for longer reads. An increasing proportion of murder mysteries now opt for the longer length, with authors like Hillerman, Burke, Smith, and James exceeding 400 pages. The advantage of length, of course, is space for more detailed character analysis and the opportunity to make plots increasingly complex and intriguing.

Character development and
ignored in the shorter novels: thos
genre. But it sometimes takes a num
the richness of the characters and t
The author builds on this continuit
Reading one police procedural out
many readers know the genre is bu
is to provide enough of each so tha
more in the next book. Short book
popular.

The popularity of the plac
partly to the success of the mystery
realistic approach taken toward th
The popularity provides a powerful
place to a large audience, many of w
author's interpretation of place—fo
these exotic places. The proof, how
themselves. A select number of pol
as examples of the effective use an
curious reader is encouraged to co
anthologies. Better yet, set aside a c
of the mystery section of your local
quality to sense of place richness, i

Place-Based Police

To date, most critiques of detectiv
place as an intriguing gimmick to l
promote other agendas, such as ecol
the case of Tony Hillerman) or femi
One is hard-pressed to even find c
component of mystery novels. In e
locale for "police novels," LeRoy I
places can add to the excitement
procedurals, the city becomes the
packed with fascination. Sometim

Robin Winks, however, do
tempt to verge on travelog. To beg

the private detective procedural, as epitomized by the works of Joseph Wambaugh. Although the main character is a hard-boiled detective, the plot revolves around unraveling the crime through the use of realistic detective procedures. A strength of each of these authors is the use of more detailed character development, like in the police procedurals, than is found in the earlier varieties of murder mysteries.

Fellow human beings to whom we can relate and empathize—this is the goal, and character development is the means to this goal. Character development, although not unique to the police procedural, is one of the keys to the success of the genre, in part, because it provides a powerful medium for realism. The typical hero of the police procedural is one with whom the reader can readily relate. The reader can associate with him or her as a fellow human being and can empathize with his or her plight. This empathy might derive from the hero's personal situation. In novels we find family men, divorcees, alcoholics or recovering alcoholics, debtors, invalids, and people struggling with complex changes in their lives. Situations are introduced in various ways throughout the novel as asides and complements to the main plot.

Character development plays a more important plot role than in earlier hard-boiled detective yarns. Often the personal lives of the characters run parallel to the story line, adding primarily to the development of the character. It is not unusual for personal lives and professional cases to become intertwined. Aspects of a character's personal life sometimes become part of the murder mystery, as in the case of James Lee Burke's Dave Robicheaux, whose wife is murdered in *Heaven's Prisoners,* or Susan Dunlap's Jill Smith, whose ex-husband's girlfriend's disappearance initiates the plot of *As a Favor.*

Also related to character development is how the protagonist solves the murder and the numerous people and situations encountered along the way. Trials and tribulations, frustrations, successes, and failures are encountered in dealing with the mundane. The manner of the investigation, of course, is presented as an integral part of the main plot. We may have never been involved in the detection of a crime, but we certainly have experienced the frustration and anger of dealing with people and circumstances that have made our lives more difficult. Either way, the key is to be able to relate to and associate with what the protagonist is experiencing—basically, the reader needs to like the hero or heroine.

The police procedural acquired legitimacy as literature with the increasing character development in novels. If earlier detectives were more caricatures than characters, the development of personality in the police procedural became an important part of bonding reader and protagonist, something unattainable in earlier detective stories. This had the added benefit of permitting a character to be used in multiple novels, developing a sometimes fanatical following.

An author had better create criminals we can relate to as well. At best, they are real people, rather than master craftsmen or sinister evil geniuses. Their motives are generally basic human motives—sex, money, power. And the criminal likewise has a history and a life that is understandable, if not condonable.

Just as it is possible to identify a number of elements common to most police procedurals, there is also a great deal of variety. When the identity of the murderer is revealed is always an intriguing part of the author's approach. We can find out whodunit early on and then spend most of the novel seeing if our police procedures will bring the culprit or culprits to justice, or we can find out the culprit at the same time as the protagonist, having spent the entire novel trying to match wits with the police in determining who the guilty party or parties might be.

Another difference in approach relates to the length. Shorter novels tend to run about 180–220 pages, and a few are even shorter. Longer novels exceed 250 pages, with some exceeding 400. This disparity gets at the heart of the discussion of the mystery story versus the novel. Originally, murder mysteries were written primarily as short stories, distributed in mystery magazines, and some even serialized in the Dickensian tradition. This continues to the present in the still popular *Ellery Queen's Mystery Magazine,* begun in 1941, and the *Alfred Hitchcock Mystery Magazine,* first published in 1956. Many of today's best-known mystery writers began with short stories in these magazines, and many continue to write short stories for them after they've become famous.

With the evolution of the genre from short stories into novels, length grew as did the popularity of the genre. Many, such as the Yellowthread Street series, the Kaminsky series, or the Dunlap series, come out at right around 180 pages, a number that translates into approximately sixty thousand words. The size is a pulp fiction style, popular with publishers because the costs of production go up dramatically with larger books. The public appetite is changing also, however, for longer reads. An increasing proportion of murder mysteries now opt for the longer length, with authors like Hillerman, Burke, Smith, and James exceeding 400 pages. The advantage of length, of course, is space for more detailed character analysis and the opportunity to make plots increasingly complex and intriguing.

Character development and intrigue, though, are by no means ignored in the shorter novels: those strengths characterize the entire genre. But it sometimes takes a number of books in the series to establish the richness of the characters and the places they inhabit and frequent. The author builds on this continuity and the addictions of the audience. Reading one police procedural out of sequence can be unsettling, but many readers know the genre is built upon repeated exposure. The key is to provide enough of each so that the reader is impatient to discover more in the next book. Short books and long have proven exceedingly popular.

The popularity of the place-based police procedural is related partly to the success of the mystery novel in general, and partly to a more realistic approach taken toward the traditional murder mystery novel. The popularity provides a powerful medium for transmitting a sense of place to a large audience, many of whom are open to and accepting of the author's interpretation of place—for many, this is their only exposure to these exotic places. The proof, however, can only be found in the novels themselves. A select number of police procedurals are used in this book as examples of the effective use and conveyance of sense of place. The curious reader is encouraged to consult any of the numerous mystery anthologies. Better yet, set aside a couple of hours to browse the shelves of the mystery section of your local bookstore: from lurid covers to prose quality to sense of place richness, it is a tactile experience. ■ ■ ■ ■ ■ ■ ■ ■

Place-Based Police Procedurals

To date, most critiques of detective fiction tend to write off the use of place as an intriguing gimmick to keep the attention of the reader or to promote other agendas, such as ecological and/or ethnic sensitivity (as in the case of Tony Hillerman) or feminism (as in the case of Susan Dunlap). One is hard-pressed to even find critiques that acknowledge place as a component of mystery novels. In explaining why cities are the primary locale for "police novels," LeRoy Lad Panek acknowledges that exotic places can add to the excitement of the plot: "In a fairly large class of procedurals, the city becomes the exotic place, full of danger, but also packed with fascination. Sometimes, this verges on travelogue."[11]

Robin Winks, however, does not simply pass off place as an attempt to verge on travelog. To begin, he supports the contention that

good mystery writing can carry within it descriptions of places that are some of the best available: "I must insist that some of the best travel writing done today occurs in the midst of spy and detective fiction."[12]

Yet, for Winks the creation of places in detective fiction plays an important role. After stressing the necessity of accuracy in descriptions of places, and the demand for careful research to ensure this accuracy, he hints at the basic concept of this book—that place matters in numerous of these novels. In referring to writers of foreign-based detective fiction: "Each of these 'exotic' writers lends verisimilitude to the plot by careful attention to local detail, so that one senses the books emerging from their landscapes."[13]

The key point applicable to place-based murder mysteries is that sense of place is not simply a crowd pleaser, nor is it used solely to advance a social agenda. It is crucial to plot development. Without place, the story line makes no sense. A crime would be agonizingly trivial. Julian Symons' comments that Sara Paretsky's novels capture the ability of murder mysteries to continue the use of realistic characters while adding place as an important component. In the Warshawski series, "[The novels] also gain a good deal from being soaked in the dangerous, but as she depicts it heartening rather than depressing, area of South Chicago."[14] This is high praise from one of Britain's most highly regarded critics.

This stylistic blueprint is also found in numerous nonmystery novels. In making "a case for popular novels as a major determinant of American place images," James Shortridge distinguishes forty-two authors between 1800 and 1950 whose novels were "place-defining," providing "concise statements of perceived regional values."[15] No one was more influential in establishing an image of the American West than Zane Grey: "Although Grey's formulaic plots have been decried by critics, all acknowledge his vivid descriptions of the vastness, wildness, and mystery of the Western landscape . . . these images remain in force today."[16]

It might be useful here to pause for a moment to make sure that we're clear on exactly what I mean by "sense of place." For the geographer, "sense of place" is part of the concept of place that includes the three major elements of "location," "locale," and "sense of place":

> *locale,* the settings in which social relations are constituted (these can be informal or institutional); *location,* the geographical area encompassing the settings for social interaction . . . and *sense of place,* the local "structure of feeling." Place, therefore,

> refers to . . . areas in which settings for the constitution of social relations are located and with which people can identify.[17]

Simply put, "sense of place" answers the question, what is it like? And the answer to that question includes all of the physical and human characteristics of the place—the physical and human landscapes, the ways in which people interact, the formal and informal institutions that structure the society, including family, church, and political and economic institutions. The focus here is that even in an increasingly integrated world, places are different and unique and that "sense of place" is about these differences, the inherently unique character of different locations in the world. Being able to convey this uniqueness effectively can be the strength of a novel and the portrayals "place-defining."

In many ways, the place-based murder mystery novel is no less "place-defining." Descriptions of place as settings for the commission, discovery, and resolution of mysteries have been the primary focus of geographic studies through the 150-year evolution of the murder mystery genre. Both Douglas McManis and Yi-Fu Tuan have explored the value of the mystery novel as an important, if underutilized, source for literary geography. McManis argues "the fact that a vast reading public is exposed to a host of geographical information from the detective-mystery genre makes it worthy of analysis for its geographical components."[18] In discussing the geographic relevance of Sherlock Holmes, Yi-Fu Tuan emphasizes more than simply the landscape descriptions of Arthur Conan Doyle.[19] Images and moods are important components of these murder mysteries, and Tuan details how they contribute significantly to plot development. Tuan is also quick to point out how the "Holmesian sagas" are analogies for Victorian England, reflecting the complex and tenuous nature of British society in the latter half of the nineteenth century, while striking chords in readers from diverse cultures for decades thereafter. The ability of literature to convey a sense of place sounds clear and powerful notes.

Much of the mystery intrigue as a function of setting and locale was maintained as the murder mystery genre underwent transformation from the classical British detective novel to the American hard-boiled detective novel to the police procedural. The relationship between sense of place and police procedurals leads to a relatively large number of series that can be classified as PLACE-BASED POLICE PROCEDURALS, the category of murder mysteries that provides the focus for this study.

Sense of place is most effectively conveyed when the plot unfolds in real-sounding places, or at least in places that could be real, as in William Marshall's Yellowthread Street, Hong Kong, police procedurals. These convey sense of place particularly well because they depict real-life situations. "The one principal factor dividing procedural from non-procedural detective fiction is not the provision of a few details for the sake of verisimilitude but the serious attempt to present or at least suggest the reality of police work."[20] The attempt to present the reality of police work carries over the attempt to create an entire novel that seems to be reality-based. With authenticity in sense of place, whatever else flows off the pages often goes unchallenged.

In describing the importance of getting places right in detective fiction, Dennis Porter suggests that "[a] crime always occurs and is solved in a place that . . . will be evoked with more or less precision."[21] Yet he sees places or landscapes primarily as backdrops, if backdrops with ideological overtones. Concentrating on the tension between the city and the countryside, for example, he interprets their symbolic value as important "as stylistic level and the type of hero."[22] Herein lies the difference. For many stories of the police procedural genre, place is merely a backdrop. In place-based police procedurals, it is the locale for the commission, discovery, and resolution of the crime, but it is much more. Place becomes an essential element in the development of the plot, without which the story falls apart. Because of the demands of greater plausibility, if not reality, in the police procedurals, place as an essential element must also be reality-based.

The need to create real-life situations with which the reader can associate puts the author firmly in the business of paying attention to both plot and character development. This has separated the police procedural from other kinds of mysteries. "Although it is the smallest of the [mystery] categories, presently the Police Procedural novels have the greatest quality of them all . . . the Police Procedural needs both character and plot, or it simply doesn't work."[23] When it works, it not only draws the reader into the plot, it not only gets the reader to identify with the characters involved, it also creates images and moods that the reader assimilates and accepts as authentic.

Four basic literary devices convey a sense of place: narrative description, dialog, iconography, and attention to detail. NARRATIVE DESCRIPTION is a basic method, although rich. We find descriptions of settings—the natural and human landscape in total—employing all of our senses: what does it look like, what does it smell like, what does it sound like, what does it taste like, and what does it feel like? And even further, what kinds of moods do these descriptions evoke? The details of these descriptions often become critical to solving the mystery. At another level, DIALOG powerfully conveys a sense of place, replacing straightforward description. It carries with it the perspectives of particular characters. It is especially powerful when the character who is speaking represents a particular segment of society, a particular and identifiable perspective, i.e., when dialog and iconography are combined. ICONOGRAPHY is often the means by which we attain our sense of a society, by using individuals and groups of individuals to represent entire segments of that society.

The final literary device of note is ATTENTION TO DETAIL. In the police procedural it is most relevant because the settings and the characters are designed to be as real as possible. This is even more necessary for "exotic" places where, more often than not, the reader has little first-hand experience or little to relate it to. This puts an increased burden on the author to get it right. Credibility is attained through the novel's attention to detail: "we see the landscape through many eyes, vicariously traveling the world. Verisimilitude requires research, and the best writers do as Gibbon did: they travel over the scenes they are to describe."[24]

When extending our need for verisimilitude from straightforward landscape descriptions to the multifaceted sense of place, the demands on the authors increase from simply "traveling over" to becoming intimately familiar with the places to be described. The allure of the police procedural is rooted in its emphasis on realism. Sense of place helps to provide the realism of the novel so that the reader will relate to the plot. The fact that the reader accepts the realism of the novel opens him or her to accept the sense of place.

NARRATIVE DESCRIPTION It begins with narrative description, which means that description serves some critical purpose in the service of the narration.[25] It may seem extraneous or unnecessary; within the context of the narrative, however, description provides a grounding in place, contributes to character development, supports a secondary agenda or subplot, provides clues for the resolution of the crime, or helps to establish the credibility of the author and the authenticity of the story.

Narrative description generally serves a number of purposes. First, it explains where and when the plot is occurring and what it's like. This gives a picture of the setting and the locale of the story. Secondary agendas can also be addressed as part of the plot setting. Social issues such as alcoholism, racism, women's rights, human rights, and environmentalism all find their way into subplots.

Recognition of these subplots captures the dual nature of place. A subplot, or several subplots, are important elements in the plot. The effective conveyance of these subplots occurs through the use of all manner of literary devices, beginning with narrative description, often in the voice of an outside, third person, lending narrative description the appearance of greater objectivity and validity. Dialog and iconography are much likelier to come across as the perspective of one particular individual in the novel.

A social issue with an interesting twist involves policewomen as lead characters. Female leads in police procedurals include P. M. Carlson's Marty Hopkins, Susan Dunlap's Jill Smith, and Julie Smith's Skip Langdon. Although all three series about women are written by women, and not a single male writer has a female protagonist, it is somewhat surprising that several other female writers have male protagonists. Most notably, the two most popular British Empire female authors of police procedurals, P. D. James and Anne Perry, have male protagonists. In the United States, although we have examples of women writing about women, we also find a continuance of the British tradition—J. A. Jance and Jean Hager are cases in point. One of the underlying social issues in the three series with female leads, unsurprisingly, is the struggle of women in a man's world. All three women serve as social icons. There the similarities end; these are three very different women. A key point, however, is that no matter how important being a woman in a man's world may be to the individual authors, it never displaces the resolution of the crime as the primary focus of the story line. If it did, we could not classify their work as police procedurals.

Key to the kinds of police procedurals examined in this study is the use of narrative description to further awareness of the commission and discovery of the crime and, more important, to provide clues for its resolution. Straightforward description builds bonds between reader and plot. The description must appeal to our senses and play upon our emotions so that we gain a stake in the outcome, so that we care about how the crime is resolved. Once engaged, we look to the description to help us outwit the protagonist, to help us figure out whodunit before that timely detail is revealed. A reader's desire to figure out the answer before the protagonist has remained constant throughout the evolution of murder mysteries and is a morphological feature common to all species of the mystery genre.

DIALOG Just as narrative description serves the "ongoing march of the narrative," so does the use of dialog, and in much the same way. A clear separation of literary devices is often impossible; they intertwine to build a good story, strengthening the bond between character (or characters) and reader and convincing the reader of the author's credibility. Dialog can help ground the reader in place, develop a character, support a secondary agenda, provide clues for the resolution of the crime, or establish the credibility of the author, the authenticity of the plot. The basic structure of the dialog has four components: who is speaking, what is said, how it is said, and who is being spoken to. Dialects, speech patterns, idioms, and local sayings all play a part in creating a believable and effective dialog.

ICONOGRAPHY We define iconography as "the identification of conventional, consciously inscribed symbols."[26] The protagonist is often an icon. In much of the murder mystery genre, iconography is employed by simply portraying the protagonist as a stereotypical representation of a particular segment of society. Police procedurals, on the other hand, employ icons to break these stereotypes. The numerous multi-ethnic detectives, such as Napoleon "Bony" Bonaparte and Mitchell Bushyhead, serve as complex characters in places traditionally oversimplified in the literature. Jill Smith in Berkeley and Arkady Renko in Moscow, and composite historic characters, such as those found in Didius Falco and Judge Dee, are only a sampling of a more complicated use of characters as icons in the police procedural.

ATTENTION TO DETAIL In details is credibility established, authenticity verified, trust secured. Mistakes in detail can render a novel unacceptable. A reader needs to believe the author has been there, has experienced the place and the culture being described, so the plot rings true. If the details suggest that the plot could not occur as described by the author, then the plot itself falls apart. If the place is wrong, then the plot could not possibly take place as written.

Not all four devices apply to each series and author. For different series that take place in the same locale, as in the cases of James Lee Burke and Julie Smith in New Orleans, or Martin Cruz Smith and Stuart

Kaminsky in Moscow, comparing and contrasting styles and approaches will be an additional focus of the analysis. Sometimes the contrasts are stronger still, as in New Orleans where the perspectives contrast a male and a female point of view.

Why Do We Read This Stuff, Anyway?

In the end, however, it all gets back to the same point. Why "connect" with the police procedurals? Why do police procedurals provide such fertile ground for eliciting a sense of place? Why are they so powerful a form of popular literature? Explanations for the popularity of detective fiction range widely. At least three kinds of interpretations of detective fiction—Marxist, Freudian, and narratological—recur in the literature. The Marxist approach views detective fiction "as a source of evidence about social constructions,"[27] wherein "literary texts are produced not only by the individuals who write them, but also by the society in which those individuals live."[28] A Freudian approach examines detective fiction psychoanalytically to discover "what is going on elsewhere."[29] From a narratological perspective, detective fiction is examined to understand structural elements of the text, uncovering how the narrative creates a story. Others would add a fourth kind of interpretation, post-modernist in approach, wherein detective fiction "involves an exploration of the experience of modernity, of what it means to be caught up in this 'maelstrom of perpetual disintegration and renewal, of struggle and contradiction, of ambiguity and anguish.' "[30]

Marxist interpretations anchor one extreme of the spectrum. Ernest Mandel sees all crime fiction as an apology for the existing social order, with crime an attack on private property, and crime fiction "turns these attacks into ideological supports of private property."[31] The crime story is "intertwined with the history of bourgeois society itself . . . perhaps because bourgeois society is, when all is said and done, a criminal society?"[32] Perhaps as fascinating as the analysis itself is the need Mandel feels to apologize for being a murder mystery fan: "Let me confess at the outset that I like to read crime stories. I used to think that they were simply escapist entertainment."[33] Once self-absolved, he provides an account of crime fiction as a reflection of the social history of bourgeois capitalism.

An even more radical approach is offered by Robert Winston and Nancy Mellerski, who see the police procedural not simply as a result or reflection of bourgeois society but as a means of control. To them, the police procedural is one of the most effective "containment structures"—it defuses

> the potential for violent transgression by fore grounding the police, a dominant Western symbol of social control . . . [T]he procedural reshapes the potentially destructive impulses of individualism into successful participation in a corporate structure, the police squad.[34]

At the other extreme of the spectrum, escapist literature is interpreted as just that—a form of escape, a pleasurable experience intensified by the author's ability to create and maintain suspense, following in the tradition of E. T. Guyown's answer to "why do we read this stuff?": "The answer comes easily. Mystery fiction is the greatest escape literature of all time."[35] Understanding its popularity had best begin with the recognition of the popularity of all escapist literature. For some, this leads to the debate over whether the escapist value means "escape *from* the problems of day to day living . . . [or] escape *into* those problems (in a pleasant way)," thus creating "a soothing effect upon the reader in reading murder mystery fiction."[36] The appeal of the police procedural includes both kinds of escapism, often simultaneously.

Others, taking less extreme approaches, also see mystery fiction as reflecting social norms and values. John Cawelti, whose *Adventure, Mystery, and Romance* is often cited as a classic study of escapist literature, sees "literary crime as an ambiguous mirror of social values, reflecting both our overt commitments to certain principles of morality and order and our hidden resentments and animosities against these principles."[37] He includes crime fiction as one kind of formulaic literature. In 1976, he was one of the first to campaign for its analysis as a legitimate genre of literature: "Formula literature is, first of all, a kind of literary art. Therefore, it can be analyzed and evaluated like any other kind of literature."[38]

For Cawelti the "mystery has become one of the most sophisticated and explicitly artful of the formulaic types."[39] The mystery story is both a reflection of the culture that created it and has influence over the culture because of the way in which society is represented and symbolized through the genre. Its strength derives from its ability to couch the escapist aspects within a context connected to reality, while supplying a resolution that is desirable and rational.

Dennis Porter echoes many of these points, emphasizing the critical role of suspense in the success of what he refers to as detective fiction. He makes a convincing argument for the importance of reality-based contexts:

> [Mainstream detective fiction] situates its actions in contemporary social reality, limits the type of crime and the methods of detection to what pass for rationally plausible, and chooses as its characters easily identifiable human or social types.[40]

Although referring specifically to landscape, Porter also makes a convincing case for the importance of place in the setting up of the crime. Place furthers the plausibility of the plot and etches its ideological significance in reflecting the social tradition in which the author works.

In providing one of the few studies directed exclusively at the police procedural genre, George Dove follows in this vein. For him, the procedural formula encompasses certain myths and legends, which serve as the foundation for the genre, and with which readers become comfortable. When a particular myth ceases to work as a plausible reflection of societal norms and values, it dies out, to be replaced by other myths.

For Dove, five conventional frames of reference are employed so consistently as to be considered basic components of the procedural formula: ordinary mortals as characters, police work as a thankless profession, the tight enclave created by police work, the importance of fickle breaks for the resolution of many crimes, and the tyranny of time. All of these speak to the need for a reality-based formula for the genre. Yet, reiterating what others have argued, the key to success is not necessarily reality, but plausibility: "Although the procedural story need not be a literal rendering, it must be at least a plausible representation."[41]

An additional motif identified by Dove in 1982 was the convention of "the happily married policeman," a convention that has undergone dramatic changes in the intervening years. Yet Dove himself forewarned of these kinds of changes over time: "When a myth ceases to have applicable value, it dies out . . . When the convention of the happily married policeman becomes unacceptable . . . we will see a change in the family patterns in the stories."[42]

We have seen just that. Currently, the happily married male police officer represents only one kind of police character type. In recent years, the genre has added widowed, divorced, gay, and female police types to the convention. What this suggests is that the demand for reality-based context elicits changes in the conventions of the formula, and that although still formulaic in nature, the components of the formula change as does the reality of the society within which the stories take place.

Within whichever context we choose to select as most appropriate,[43] the police procedural gathers much of its allure from its emphasis on realism. Although the kinds of murder depicted in the typical murder mystery are few and far between in real-life police work, as Hillary Waugh points out, they can and do occur in real life. More to the point, the fictional characters, the procedures, and the situations are depicted realistically. They are believable and human, much easier to relate to, with a few exceptions, than the bigger-than-life, hard-boiled American private eye, or the superhuman, amateur British sleuth.

This is why place is such an important component of the police procedural. If the reader is to believe the author and buy into the realism of the novel, then a realistic, believable sense of place is an essential element in creating realism. If a key to success is to focus on the mundane experiences of everyday life, then the rhythm of everyday life as captured in sense of place is the baseline for believability. The author must express this rhythm in descriptions of setting and locale, of manners and customs of the culture and society, and of interaction between characters. The stronger the sense of place and the realism of the novel, the stronger the ties between the reader and the characters. In fact, it's a two-way street. Sense of place helps to provide the realism of the novel so that the reader will relate to the plot. Accepting the realism of the novel opens the reader to accept the sense of place as created by the author.

Murder in America
Young
Jance
Chapter 3
Dunlap
Carlson
Hillerman
Hager
Burke
Lindsey
Smith
Taibo

It is only fitting that the investigation of place-based police procedurals begins in America, where the police procedural was invented and turned into a literary art form. Unfortunately for true devotees of the procedural genre, our requirements for being place-based necessarily exclude a number of authors largely responsible for the success of the police procedural, namely Hillary Waugh, K. C. Constantine, and Ed McBain. Although their work will not be examined in detail, it would be criminal not to acknowledge their contributions to police procedurals.

Hillary Waugh is often credited as the instigator of police procedurals as a popular genre of escapist literature. *Last Seen Wearing*, published in 1952, began a literary trend that spread worldwide. Not only is Waugh a prolific and popular writer of police procedurals, he is often cited as a critic of the genre. Although Waugh is a major critical commentator of the mystery genre, and although his novels are excellent police procedurals, they lack a strong sense of place.

For excellent examples of detailed and real-life police work, any one of Waugh's Chief Fred Fellows mysteries offers an example of the craft of a true artist of the procedural. Unfortunately, there is little sense of Stockford, Connecticut. Even though Stockford is a fictitious town, Waugh's work is almost entirely without a sense of a small northeastern town as setting for the mysteries. The crimes could have occurred in a small generic town anywhere in the United States. More importantly, Stockford as place fails to figure significantly in the commission, detection, or resolution of the crimes.

The same can be said for K. C. Constantine, who is also an immensely popular author of police procedurals. The murders solved by Rocksburg Police Chief Mario Balzic are also set in a small town, this time in western Pennsylvania. One of the defining features of the Balzic series is character development. Dealing with crime sometimes takes a backseat to Mario's personal life as with the death of his mother in *Sunshine Enemies,* and sometimes there isn't even a murder as in *Bottom Liner Blues.* These are place-agnostic books, with a lack of any sense of where the series takes place—officially the fictitious Rocksburg, Pennsylvania. Other than sampling occasional insights into what life is like in a predominantly Italian-Catholic factory town, there is nothing about the town itself or surrounding areas. This is a series that could happen anywhere, and the place is not an important plot element.

Ed McBain, with his 87th Precinct series, is the acknowledged master of the American police procedural. Although the 87th Precinct is located in the fictional city, Isola, everyone knows it is New York City. Since McBain is a master of all of the essential elements of the police procedural, it is easy to see why the series numbers forty-six books: "there ain't no way you can turn a murder mystery into a silk purse. That's because the minute somebody sticks a knife in somebody else, all attention focuses on the victim, and all you want to know is whodunit" (*Romance,* p. 23).

This excerpt from a recent 87th Precinct novel gives the source of the power in murder mysteries. Once a murder is committed, attention is set on solving the murder and uncovering the murderer. That is the nature of the crime—a violation sufficient to rivet attention. Because of this focus, the author is able to get away with all sorts of subplots and hidden, or at least secondary, agendas. In addition to capturing the essence of the murder mystery so eloquently, Ed McBain's novels are characterized by the classic strengths of the police procedural genre—attention to police procedural details, police working as a team, detailed character development—all wrapped in a story line that could have happened in real life.

One area that is not particularly strong in the 87th Precinct series, however, is the sense of place. Although the reader can readily identify the descriptions of the city as being replicas of New York City, one doesn't get a real "feel" for the place, other than it's a large metropolis in the United States. It is probably because McBain has attempted to create a more generic urban place for the setting of his mysteries that the sense of place comes across only weakly. Like Hill Street in the 1980s TV show, place is purposefully generic.

The elimination of these three popular series notwithstanding, there is no real limit to American series. The ten series selected provide a cross-section. The first author selected is Tony Hillerman, who many acknowledge as the model for cultural murder mysteries, if for no other reason than he has been so successful in promoting place as an essential part of his mysteries. He clearly defines Navajo culture and the American Southwest for millions of readers.

Although not as commercially successful as Hillerman, Jean Hager provides another Native American approach by setting her police procedurals in Cherokee country, Oklahoma. It has been suggested that Hager does for the Cherokee what Hillerman does for the Navajo. Although that's probably an overstatement, Hager's series is worthy of examination.

Other regions have their definers as well. New Orleans is the locale for series authored by James Lee Burke and Julie Smith, two very different approaches to the "Big Easy" as place, not the least of which is the appearance of the first female protagonist. David Lindsey's series is based out of Houston, but he uses a number of excuses to get his hero out of town and to expose readers to Mexico and Guatemala. The Midwest finds representation through the writing of P. M. Carlson, whose female cop calls southern Indiana home. J. A. Jance sets murders in the serenity of her American Northwest, while Susan Dunlap lures us to the anything-but-serene Berkeley to solve her mysteries through the exploits of another female cop. Venturing across the border to the north, Scott Young introduces eastern and northern Canada through the eyes of another Native American, an Inuit protagonist this time, and venturing south of the border to Mexico, Paco Taibo II provides the first foreign-language murder mysteries covered here, several of which have been translated into English.

Just to keep track of the numbers, our ten authors include five female authors but only three female lead characters; three Native American lead characters but no Native American authors; two series based outside of the United States, although one other series takes off to Latin America on occasion. Places for dead bodies include Canada, Mexico, and the American Northwest, West, Southwest, Midwest, and South.

The Navajo Country of Tony Hillerman

[Joe Leaphorn] used the map in his endless hunt for patterns, sequences, order —something that would bring a semblance of Navajo hohzho *to the chaos of crime and violence.* (Coyote Waits, P. 150)

TONY HILLERMAN IS THE DEAN of the place-based police procedural in the United States, if not the world. His popularity and success are evidence that cultural context, done well, not only produces first-rate, award-winning novels, it also sells big time. In addition to the Joe Leaphorn/Jim Chee series consistently reaching the best-seller list, Hillerman has won the recognition of his peers for being a master at his craft. He has received the most prestigious mystery writing awards, including the Mystery Writers of America's Edgar and Grand Master awards. As significant, he has won recognition from Native American groups as well, receiving the Center for American Indians' Ambassador Award and the Navajo Tribe Special Friend Award. As a result of his success and popularity, Tony Hillerman has made a major impact on a generation of readers' perceptions of Native Southwest American culture and the landscape of the American Southwest, a responsibility he doesn't take lightly.

Just as any good police procedural requires in-depth research on the real-life workings of the police department, the place-based police procedural requires meticulous, if not first-hand, knowledge of the region used for the setting. Most authors gain their knowledge of police work through hours of research, and their familiarity with the region from living or spending a great deal of time there. For Hillerman, the task is even more challenging because he must not only get the police procedures and the landscape right, he must also be accurate about the Native American cultures, making them understandable to the non-Native reader and acceptable to and respectful of the Native reader. Hillerman succeeds on both counts. There are some inaccuracies, but these inaccuracies are not offensive and do not diminish the impact of his sympathetic treatment of Native culture. Hillerman presents Native practices and beliefs at a level that pays respect to the culture while educating non-Natives about the seriousness and importance of these traditions.

The heroes of the series are two Navajo tribal policemen, Joe Leaphorn and Jim Chee, each representing quite different kinds of Navajo—here begins the use of characters as icons. Lieutenant Joe Leaphorn, the older of the two, represents an unusual blend of Navajo and Anglo cultural influences. During his years at Arizona State University, Joe was well on his way to becoming fully assimilated into Anglo-American culture. His return to the reservation was precipitated by his marriage to a nineteen-year-old Navajo, Emma, who would be his wife for thirty years, until her untimely death is revealed at the beginning of *A Thief of Time.* Emma served as Joe's main contact with the Navajo Way, an influence that endures after her death. Although he has lost his belief in Navajo spirits and superstitions, Joe retains a great deal of pride for the Navajo nation and a deep respect for its people, even though he is often impatient with many of the Navajo traditions and customs.

Jim Chee, on the other hand, represents a different kind of icon. A Navajo tribal police sergeant through most of the series, he too has been influenced rather dramatically by the conflict between the two cultures—Navajo and Anglo. Like Leaphorn, Chee is a college man, educated at the University of New Mexico, where he adapted well to the ways of the white man. Yet, unlike Leaphorn, his return to the reservation and the Navajo Way is by choice, even to the point of studying to be a Navajo Singer (Medicine Man). He knows and practices Navajo prayers, songs, and blessings. This knowledge serves him well in a number of the cases. The major internal conflicts he endures concern the divergent pulls of Navajo and Anglo cultures as well as the pulls of his work as a policeman and his desire to be a Navajo Medicine Man.

If educating readers about Native Southwest American cultures is one of the underlying themes of the series, Hillerman's perspective on the need for unity between society and the natural environment is a complementary theme. This is an aspect of Native American culture he chooses to emphasize. Yet Hillerman feeds it into the plot without being preachy. The importance of the unity of society and the environment is presented as a fact that must be appreciated to understand the plot and the resolution of the crime. Its infusion into the novel works because it never overpowers or diminishes the main purpose of the novel—providing an entertaining whodunit. Lessons on environmental sensitivity and Native Southwest American culture are crucial subplots but not the primary focus.

Hillerman gets his geography right. Some books in the series contain maps; where maps are not provided, readers can refer to their

"Indian Country" map to follow Leaphorn and Chee's movements. The reader is alerted when fictitious places are to be introduced, blending the fictitious and the real. His fictitious places fit well into the story line because they could exist in the real landscape.

It all begins, of course, with the description of the Southwest American landscape. This is another of the strengths of the Hillerman series. Hillerman might be referred to as the "James Michener of the place-based police procedural." His novels offer vivid and penetrating images of the great American Southwest, drawing readers

> out across the gentle slope that fell away from Tesihim Butte and then rose gradually toward the sharp dark outline of Nipple Butte to the west. The sage was gray and silver with autumn, the late afternoon sun laced it with slanting shadows, and everywhere there was the yellow of blooming snakeweed and the purple of the asters. (*Talking God,* p. 29)

Often straightforward landscape description is intertwined with Indian folklore and mythology. These are not just striking desert landscapes; they are also the sources of much of the Native cultural traditions.

> We tend to think of heaven as being up in the sky. The Zunis also have a geographical concept for it, because of the nature of their mythology. Do you know that myth? (*Dance Hall of the Dead,* p. 144)

> Across the highway from it, slanting sunlight illuminated the ragged black form of Barber Peak, a volcanic throat to geologists, a meeting place for witches in local lore. (*Coyote Waits,* p. 102)

Mythology serves two purposes. On the one hand, it helps to depict more vividly Navajo culture, an important underlying theme in the Hillerman series. On the other hand, it provides critical clues to the mystery. In the particular case of these lava formations, discovering who was breaking the taboo, and why, was essential to uncovering the murderer of Police Officer Delbert Nez. A third service is provided by the description—securing the credibility and authenticity of the story.

The series pursues a number of variants with the two main characters. Early novels used only one or the other. Not until *Skinwalkers* would Hillerman bring Jim Chee and Joe Leaphorn together in the same novel. Sometimes they work on two totally separate cases. Other times they work on the same case from different angles and for different reasons. As well sometimes, they begin working on seemingly different cases that end up being the same case by the end of the novel. All scenarios work well.

Throughout the Hillerman series, homage is paid often to the value of geography and geographic principles. Looking for patterns on the landscape, for example, is one skill that makes Joe Leaphorn such an effective policeman.

> Joe Leaphorn still remembered not just the words but the old man's [Haskie Jim] face when he said them: "I think from where we stand the rain seems random. If we could stand somewhere else, we would see order in it."
>
> After he had thought about the meaning in that, Leaphorn had looked for order in everything. And he usually found it. (*Coyote Waits,* p. 257)

The search for patterns and order is most geographically depicted in Leaphorn's use of maps, or at least one legendary map.

> Leaphorn's map was known throughout the tribal police—a symbol of his eccentricity . . . a common "Indian Country" map published by the Auto Club of Southern California and popular for its large scale and its accurate details. What drew attention to Leaphorn's map was the way he used it. (*Skinwalkers,* p. 19)

And use it, he did. Numerous pins, each color depicting a different kind of crime, hundreds of notes about the physical landscape, locations of ruins and historical sites, temporary settlements and camps, and important places in Navajo mythology adorn a map that aids in his never-ending search for patterns, patterns that often provide insight into the crimes he investigates.

As well, continuing antagonisms between white society and the Native Americans is a recurring subplot throughout the series. This antagonism serves as the setting for *A Thief of Time,* Hillerman's personal favorite in the series and thought to be one of his best novels, in which competing anthropologists are illegally searching for ancient Anasazi pottery, thus defiling ancient ruins and "[s]tirring up Anasazi ghosts."

Not only does this novel include an intricately woven plot that involves both Leaphorn and Chee, it takes a serious look at Native American and Anglo religious beliefs and how they come into conflict in contemporary Navajo culture. Hillerman is sensitive to the fact that these kinds of activities are a constant threat to sacred Navajo grounds and takes care to hide the location of the actual ruins: "AUTHOR'S NOTE: While most of the places in this volume are real, Many Ruins Canyon has had its name changed and its location tinkered with to protect its unvandalized cliff ruins" (Preface to *A Thief of Time*).

Differences within Navajo culture are explored as well. In the person of Janet Pete, for example, Hillerman underscores the impact of Anglo culture on someone who had been raised off the reservation.

> [B]y Navajo standards, such an interruption was rude. One let a speaker finish, and then waited to make sure he was indeed finished, before one spoke. But then Janet Pete was really Navajo only by blood and birth. She hadn't been raised on the Reservation in the Navajo Way. (*Coyote Waits*, p. 143)

In discussing the route over Washington Pass in the heart of Navajo land, there is a reminder that something as simple as place names serve as constant reminders of the white conquest of Native American lands and, at times, the insensitivity of the dominant culture.

> Why does [the white man] honor the man who was our worst enemy and rub our noses in it? . . . Why take such a bastard and name a mountain pass right in the middle of your country after him? Is that just the product of ignorance? Or is it done as a gesture of contempt? (*Coyote Waits*, p. 204)

Tony Hillerman provides vivid senses of place through the use of description, dialog, iconography, and attention to detail. All four literary devices are employed masterfully in producing not only a thriller of a crime series but also an impression of the Native American Southwest as place. In the end, the reader has a new-found respect for native cultures and feels as if he or she has been a part of these cultures for a short while. This is one of the strengths of the Hillerman style. It is also why his writing represents some of the best examples of the place-based police procedural. ▪▪▪▪▪▪▪▪▪▪▪▪▪▪▪▪▪▪▪▪▪▪▪▪▪▪▪▪▪▪▪▪

The Cherokee Country of Jean Hager

Mitch was mesmerized by the story with its sacred regard for the number seven, the theme found in the myths and legends of many cultures.
(The Grandfather Medicine, P. 186)

LIKE HILLERMAN, Jean Hager's Cherokee-based novels, with a great deal of sensitivity, bring to the attention of non-Natives a number of issues important to the Cherokee nation. Yet there parallels begin to fade. Unlike Hillerman's, one of Hager's series set among the Cherokee is not a police procedural. The more popular series, which follows the exploits of Molly Bearpaw, investigator for the Native American Advocacy League, is not truly a police procedural, but it does deserve brief mention as part of the overall offering of Cherokee murder mysteries.

From the perspective of sense of place, the Molly Bearpaw novels fall short, missing an excellent opportunity to provide us with images of Cherokee country and a greater understanding of Cherokee culture. The series utilizes only weak landscape description; her handling of Cherokee culture improves over the course of the series, but it is not germane to the mystery, other than providing the *raison d'être* for Molly's involvement in the case, and loses its impact on the reader. This appears to be the main difference between Hager and Hillerman. The reader does not feel as if he or she truly experienced Cherokee culture by the end of the mystery.

Hager's true police procedural, although not as popular as the Bearpaw series, does a better job using sense of place. Mitchell Bushyhead is chief of police in the fictitious town of Buckskin, located about an hour's drive from the real city of Tahlequah in eastern Oklahoma. Although there are references to the physical landscape throughout the novels, the strength of the Bushyhead series is in the depiction of Cherokee culture, which is essential to plot development.

The prologue to *The Grandfather Medicine* begins with the enactment of a medicine ceremony that sets the stage for the discovery of the first murder. The discovery of why and for whom the ceremony was performed helps in the resolution of the crime, as do the revelations of the Medicine Man, Crying Wolf. More than that, the antagonisms between the three cultures of Buckskin—full-blood, half-breed, and white—constitute an undercurrent for the plot that reveals the complexities of Cherokee culture. As an iconographic figure, Mitch is the bridge between the Cherokee and the white cultures.

In the Bushyhead series, the dominant subplot is the underlying tension between Cherokee and Anglo cultures. The series shares a number of similarities with Hillerman's: a sensitive portrayal of Native culture and traditions, often intertwining doses of Native American history as background, the use of characters as icons, and attention to detail. Its descriptions of landscapes are weaker, sacrificed for an emphasis on culture; the strength of the series is also its weakness.

Evidence of a culture in conflict with the Anglo world is Chief Mitchell Bushyhead, a recently widowed, single parent with a teenage daughter. Trials and tribulations of a single parent are interjected, paralleling the early stages of a blossoming romance. A "half-breed" Cherokee, he finds himself not completely part of either culture.

> Mitch was still considered an outsider by many of the full-bloods in Cherokee County. It wasn't his half breed status that had put a wall between him and them; he'd encountered little racial prejudice among Cherokees. He was distrusted because he hadn't been raised in the Cherokee way and therefore couldn't be expected to understand anything having to do with their Indianness. (*The Grandfather Medicine,* p. 78)

This sounds familiar—similar to Hillerman's description of Janet Pete in *Coyote Waits.* Mitch's "half-breed" status is set within the context of the history of Cherokee integration into the dominant Anglo culture.

> The Cherokees were . . . among the first to embrace the white man's culture. True, they had done so out of necessity, seeing it as the only way to preserve their tribal identity and remain a separate nation. In spite of their efforts . . . they hadn't been a nation with their own territory since statehood in 1907. As a group Cherokees had probably intermarried with whites more extensively than any other North American Indian tribe. There remained pockets of conservatives who continued to fight assimilation, but even after living in this rather isolated corner of Cherokee County for ten years, it still surprised Mitch how many Cherokees continued to put their faith in Cherokee medi-

> cine. He had never felt any affinity with that kind of Cherokee; and he hadn't expected to come away from the meeting with Crying Wolf with such a strong feeling of respect for the old man and his beliefs. (*The Grandfather Medicine,* pp. 200–201)

Not only does the reader get a brief lesson on the history of Cherokee-American government relations, as interpreted through the eyes of the Cherokee, the reader is also appraised that Mitch's assimilation into the dominant Anglo culture has been quite effective. This passage serves as a reminder of the importance of the traditional beliefs of his Cherokee ancestors. Because we also are reminded of the importance of traditional Cherokee beliefs, this epitomizes the primary focus of Hager's use of sense of place—the continuing struggle within the Cherokee community itself between holding onto traditional Cherokee culture and assimilating into the culture of the dominant Anglo society. Although not always essential for solving the crime, Hager uses this struggle of cultures as a setting for murder. In *Night Walker,* the murder is committed at a resort lodge, built on an ancient Indian graveyard. One can never totally dismiss the possibility that the motive is related to the desecration of this sacred land. It certainly provides a venue for exploring the legacy of Cherokee religious values and the unceasing onslaught of progress in Anglo society. The confrontation of cultures is examined at the level of society as a whole as well as on a personal level.

In *The Fire Carrier,* Mitch begins to come back in touch with his Indian roots through his association with Dr. Rhea Vann, who recently returned to Buckskin to run the Cherokee Nation's family clinic, and whose grandfather is Crying Wolf, who figured prominently in *The Grandfather Medicine.* Rhea and Crying Wolf convince Mitch to attend a stomp dance—an age-old Cherokee social gathering intended to build a sense of community and to keep alive Cherokee legends and traditions. As Mitch explains to friends he meets at the gathering, he received an invitation because Crying Wolf "thinks I need to be educated in the way of the Cherokee" (*The Fire Carrier,* p. 133). Mitch spends the rest of the night becoming intrigued about his Cherokee roots, of which he is relatively ignorant. After the stomp dance, Mitch solves the murder, leaving us to wonder where his new-found interest in his Cherokee roots, not to mention his new-found relationship with Rhea Vann, might be heading. Such is the lure of the series. Now the next installment is eagerly anticipated, and not just to get at another well-written mystery plot but to discover how these issues have been resolved. ■■■■■■■■■■■■■

The New Orleans of James Lee Burke

The street was loud with music from the bars and strip houses, and the sidewalks were filled with tourists, drunks, and street people who were trying to hold on to their last little piece of American geography.
(The Neon Rain, P. 138)

JAMES LEE BURKE'S moving and compelling images of New Orleans and the bayou country of southern Louisiana provide the context for the exploits of Dave Robicheaux, who starts out as a member of the NOPD and then retires from the force after the first novel to solve murders as a nonprofessional in the shattered peace and tranquility of the Louisiana bayou. Dave, however, cannot resist the call to duty, and by the third novel, he takes a position as a sheriff in New Iberia, about 175 miles due west of New Orleans. Yet, even after setting up shop in New Iberia, Dave spends much of his time in New Orleans, which is the primary setting for *Dixie City Jam,* which ends with a rather poetic and nostalgic description of the city:

> The music rose into the sky until it seemed to fuse with the gentle and pervasive light spreading far beyond the racetrack, over oak-lined streets, paintless wood houses with galleries and green window shutters, elevated highways, the Superdome, the streetcars and palm-dotted neutral ground of Canal, the scrolled iron balconies, colonnades, and brick chimneys in the Quarter, Jackson Square and the spires of St. Louis Cathedral, the Café du Monde, the wide, mud-churned sweep of the Mississippi, the shining vastness of the wetlands to the south, and eventually the Gulf of Mexico, where later the moon would rise like an enormous pearl that had been dipped in a glass of burgundy. (*Dixie City Jam,* p. 494)

The New Orleans Chamber of Commerce would have been hard-pressed to have come up with a better description for luring visitors to the "Big Easy." Yet, placed at the end of five hundred pages of a rather complex and intense story line, in which Dave, his colleagues, and friends barely survive, saving New Orleans from the machinations of some pretty surly and downright mean villains, this description does not seem out of place. The New Orleans of legend through the exploits of Dave Robicheaux is seen to have been preserved, remaining unscathed as the jewel of the American South.

As the Dave Robicheaux series progresses, James Lee Burke increases the length of the novels from the 280 pages of *The Neon Rain* to the 500 pages of *Burning Angel.* Yet, no matter the length, he still holds the reader with a gripping plot. Several components contribute to the success of the series and to its power as a place-based police procedural.

In Dave Robicheaux, James Lee Burke has a likeable, ethical hero. Robicheaux quits the NOPD because he can't deal with what he sees as corruption within the department. As is often the case, Dave becomes more than just a police officer. Many of the pages focus on his private life and the problems that he must deal with personally in order to carry on professionally. Dave Robicheaux carries a number of classic post-Vietnam demons. He is a divorced, alcoholic Vietnam vet. As a result, we find numerous commentaries on alcoholism and Vietnam; yet, surprisingly, he seems to have plenty of booze available for guests in his houseboat—strange for an alcoholic that struggles the way he does. Later in the series, his wife will fight a bout with alcoholism as well.

In *The Neon Rain* and *Dixie City Jam,* much of the plot centers on Dave himself. He is not just a cop doing his job to solve the murder. Rather, he and later his family and friends are involved in the murders, even to the point of being targets and suspects. This technique is used successfully by a number of authors. Readers are endeared to main characters who appear vulnerable and real. The most immediate and tragic involvement of his family occurs while he is not a cop but has relocated to New Iberia to run a bait shop on the bayou. In *Heaven's Prisoners,* his interest in murder leads to the murder of his wife, Annie.

There is widespread fascination with the exotic nature of New Orleans to begin with. The "Big Easy" epitomizes the sensual and the forbidden, a place of unrivaled food, entertainment, and culture—all to excess. Burke uses these attributes to advantage by providing lively images of New Orleans and southern Louisiana, both in description and in mood, with a bit of history thrown in for good measure.

Particularly effective, and not all that surprising given its world-renown status, is the use of New Orleans food, drink, and music. A mood that helps convey a sense of place is provided by the intrusion of jazz and zydeco, immediately associated with New Orleans. We spend endless hours eating at various kinds of restaurants, from the highly classy to the local boathouse. Oysters on the half shell, gumbo, po' boy sandwiches, Jax beer, and indigenous dishes loiter among the pages of the novels. You smell and taste them. Salty air and the smells and sounds associated with New Orleans remind everyone of its location as a major port city.

There is a continuing antagonism between the beauty and excitement of the city and its seamier, slimy underbelly. Burke juxtaposes the beauty and serenity of the surrounding bayou region with the decaying city that has been overridden by crime, corruption, and pollution. Yet, throughout all of this balancing, there is no question that Burke has a love affair with New Orleans. Only from an author who truly loves the place he describes can the reader accept the interpretations as real and legitimate. Our first introduction to the French Quarter carries this interpretation:

> As always, the Quarter smelled to me like the small Creole town on Bayou Teche where I was born: the watermelons, cantaloupes, and strawberries stacked in crates under the scrolled colonnades . . . the cool, dank smell of old brick in the alleyways.
>
> [B]ut the majority of Vieux Carre residents [today] were transvestites, junkies, winos, prostitutes, hustlers of every stripe, and burnt-out acidheads and street people left over from the 1960s. (*The Neon Rain*, p. 12)

Seven novels and a decade later, Burke continues to play on this antagonism.

> Morning was always the best time to walk in the Quarter. The streets were still in deep shadow, and the water from the previous night's rain leaked from the wood shutters down the pastel sides of the buildings, and you could smell coffee and fresh-baked bread in the small grocery stores and the dank, cool odor of wild spearmint and old bricks in the passageways.
>
> But it wasn't all a poem. There was another reality there, too: the smell of urine in doorways, left nightly by the homeless and the psychotic, and the broken fragments of tiny ten-dollar cocaine vials that glinted in the gutters like rats' teeth. (*Dixie City Jam*, p. 19)

Another technique used by Burke, and one particularly appropriate to the Gulf region, is descriptions of the weather. Southern Louisiana's subtropical climate impacts each story. Here is the kind of detail that reminds us that we are being guided through the plot by someone who knows the effects of heat and humidity. And the heavy rain showers and thunderstorms, a reality of bayou life, have a propensity to hamper and/or help both the heroes and the villains.

It's not just sense of place that pervades the pages of the Robicheaux series. Often Burke explicitly refers to geography in diverse contexts, giving a flavor for the importance of geographical relationships often taken for granted: "my wife had taken up residence in that special piece of geography where the snakes hang in fat loops from the trees and a tiger with electrified stripes lights your way to his lair . . . What were the ongoing connections in the Buchalter case? Music, and now geography" (*Dixie City Jam*, pp. 215 and 436).

The geography of New Orleans and southern Louisiana is critical to the resolution of all of the murders in the Dave Robicheaux series. Spatial relationships, patterns, and just plain physical geography are all important components of the mystery. They also enable one to better understand this most special of American places. ▪▪▪▪▪▪▪▪▪▪▪▪

The New Orleans of Julie Smith

New Orleans, though technically a city,
is more like a nation unto itself; though legally
a piece of America, is Caribbean in its soul,
as exotic an adventure as exists
short of navigating the Amazon.
(Jazz Funeral, pp. 1–2)

IN NEW ORLEANS we have a second interpretation of the same place, this one through the perspective of a young female cop, providing a counterbalance to Burke's Dave Robicheaux. In contrast to the good ol' Cajun boy aura of Burke's novels, Julie Smith creates Skip Langdon—not just a policewoman, but one who has been raised in New

Orleans high society, a fact used to explain her rather unconventional involvement in homicide investigations while she is still a rookie beat cop. Later, she will be promoted to homicide detective.

Unlike most female detective series, in which an important subplot is the struggle of a woman in a man's world (although traces of that are evident in this series as well), Skip Langdon's primary struggle is with her upbringing as a proper, genteel Southern belle, belying her stature of being six feet tall and overweight. Choosing police work as a career brands her a "black sheep" in her socially prominent family. She carries a cynicism about the upper classes, particularly toward the traditional "perfect-hostess-in-any-circumstances" role of women in Southern society.

Skip believes that she is not alone in her antagonism toward her upbringing as a traditional Southern socialite.

> After a moment it came to [Skip]; it was usually Southern women who were treacherous. And not all of them, either, only the wildly unhappy ones who'd gotten trapped in the steel-magnolia syndrome and resented it in bilious undercurrents that made their families miserable and erupted at funerals and weddings—any inappropriate time guaranteed to embarrass everyone present. (*The Axeman's Jazz,* p. 177)

Unlike the Burke approach to narrative description, Julie Smith's descriptions of New Orleans tend to be commentaries on the relationship between New Orleans and those who live there.

> New Orleans could wreck your liver and poison your blood. It could destroy you financially. It could shun you or embrace you, teach you tricks of the heart you thought Tennessee Williams was just kidding about. And in August it could break your spirit. (*The Axeman's Jazz,* p. 1)

Smith's concern with emphasizing the duality of place and the contradictions that epitomize New Orleans recurs throughout the series. It is clear that she interprets this duality and perversity as not only unique, but seductive in its own way: "Even those who have lived there for years and those who were born there and those who have spent their last twenty years drunk in the gutter can feel the strangeness of the city, its seductive perversity" (*The Kindness of Strangers,* p. 1).

series and one of its great appeals. In *In the Lake of the Moon,*
er directly involves Stuart Haydon and the history of his family.
ılt, we learn much about Stuart Haydon and his father, a bit
s wife, Nina, and his maid Gabriella, and much less about his
nd her side of the family. Although Haydon is independently
the reader is not put off by it because Haydon is a real person
rough a quest for the truth about his father's past, revealing a
is father he knew nothing about. The story line is universal; it
ppen to you!
ven when the story line is not directly about Haydon and his
naracter development is quite detailed. The actual murders don't
e until late in the novels, and by then we're pretty sure who's
be murdered. The only questions are how, why, and who else.
es, the character development is used effectively by Lindsey to
ıs with a history of the places involved.
n one instance, Lindsey relates the history of Mexico by focus-
ıe history of one of its streets.

icenciado Verdad was a short street with a long history. In the
arly sixteenth century when Hernán Cortés with his Spanish
oldiers and Catholic priests crossed the long causeway and en-
ered the Aztec capital of Tenochtitlan. . . .

After the Spaniards conquered the Aztec empire in 1521,
hey razed the ancient city . . . and then modern Mexico replaced
he colony, but the geographic center of each of these empires
emained the same, the heart of Mexico City. The narrow, alley-
ike street of Licenciado Verdad had seen it all. (*In the Lake of
he Moon,* p. 301)

Although native folklore and mythology do not play the role they
llerman or Hager, there is a heavy dose of native Central Ameri-
ure, used basically to help explain the nature of Mexican society

"So, what did the Indians learn about life, about themselves? At
the very core of the later Mesoamerican religions is the concept
of duality . . . the unity of basically opposed principles. Life and
death, light and darkness, good and evil, up and down, yes and
no. Each god was more than one god; sometimes he was many,

Although different in approach and perspective, both of these series share crimes and resolutions steeped with the culture and geography of New Orleans and southern Louisiana. If Robicheaux is constantly involved with pimps, prostitutes, drug dealers, and other assorted lowlifes, Langdon is usually entrenched with the wealthy, the powerful, and the prominent. Together, the two series complement each other nicely.

The Houston and Latin America of David Lindsey

Mexico City had become a grotesquerie.
And yet, as if still clinging to the dualism that
pervaded the ancient Aztec theology that had flourished
in the once-beautiful Valley of Mexico
six hundred years earlier, the city had another face.
(In the Lake of the Moon, p. 164)

FOUR HUNDRED MILES to the west of New Orleans, David Lindsey gives us Homicide Detective Stuart Haydon of the Houston Police Department. Fortunately for the reader, Stuart often gets involved in cases that take him out of Houston to Latin America. It requires a stretch of the imagination to define these novels as police procedurals because most of the plot occurs out of Stuart's jurisdiction. But the sense of place is so poignant, and Latin America is sufficiently underrepresented in the police procedural literature, that the Haydon series adds a welcome component to our regional ramblings.

First, however, we learn of Houston itself—often quick glimpses from time to time. In setting the weather and mood of the city in fall, Haydon notes that Houston

> was beginning to smell fusty, and the trunks of the trees were sprouting a fine, linty pelt of Kelly-green lichen. Every color exposed to the weather was two shades darker from the incessant

> moisture . . . The bayous had risen and stayed up, and some parts of the city flooded every few days. (*In the Lake of the Moon*, p. 6)

A favorite Lindsey subplot is the contrast between the rich and the poor, employing images of neighborhoods to bring home the contrast. It also provides an opportunity to share displeasure at the development of downtown Houston over the years.

> Haydon's route home did not require him to travel the freeways that embraced Houston like tentacles of sargasso reaching in from the Gulf of Mexico to strangle the city with congested veins of simmering cement. Instead he turned from the blistering downtown canyons of steel and glass. (*A Cold Mind*, p. 47)

The tour of Houston takes us to some pretty exclusive areas, such as the St. Remy health spa.

> Haydon stopped at the guardhouse, identified himself, and pulled into the narrow lane leading through the towering pinewoods that formed an arboretum over the entire estate. For a while the lane followed alongside the jogging trail, which was covered with Superturf and twisted through the forest like a mossy carpet. Occasionally the pines would open to a meadow, with cleanly mown emerald grass. (*A Cold Mind*, p. 137)

A short while later, Haydon finds himself in a very different Houston: "He squinted back in the glaring sun at the Marsdon Court housing projects. Their cement-block dormitory-style architecture set the tone of hopelessness that smothered the barrio neighborhood like a stultifying tropical heat wave" (*A Cold Mind*, p. 146).

Much of the focus of this series centers geographically on Latin America—Mexico City in *In the Lake of the Moon* and Guatemala in *Body of Truth*. Quite different senses of place emerge from each of these novels. Not coincidentally, his ramblings outside of the United States take place after the death of his partner, a time when not only is he on his own, and thus capable of these unusual detours out-of-country, but also a time when he goes through a great deal of self-examination, where physical journey parallels psychological journey.

The Mexico City of *In the Lake of the Moon* is presented ambiguously, offered as a love-hate relationship between Haydon and the city.

Lindsey intertwines outright description,
thology and folklore—quite detailed and qu
lifelong knowledge of this sprawling, out-
alternating affection and abhorrence for a
Lake of the Moon, p. 163).

One can sense this "dualism" in t
captures the layout of the city.

> The city's layout owed more to ro
> than it did to logic. . . . It was a nigl
> the metaphor of the city as labyrin
> ing. Sometimes it bordered on the
> *Moon*, pp. 166–167)

The Guatemala of *Body of Truth* do
an interpretation. The sense of place portra
political, combining a sense of terrorism an
American duplicity in an unambiguously r

One of Haydon's contacts in Guat
plains the situation quite simply.

> "One: the army controls everythi
> wealthy elite that holds most of the
> trols the economic fortunes of the
> upholds the inequitable situation, a
> this mercantile elite there is an eas
> losers are 90 percent of the populati
> to change. The civilian 'governme
> *Truth*, p. 117)

As adjunct to the overall political si
reminded throughout the novel that tortu
daily routine, and that impending visits by g
squads are a way of life for many Guatem
of the extreme poverty and despicable livin
malans. The support of the U.S. governme
bring democracy to Guatemala cannot be a
of the deaths of a number of people are acce
context.

Detailed character development is

Haydo
the mu
As a r
about
mothe
wealth
going
side of
could

family,
take p
going t
In all c
provid

ing on

do in H
can cul
today.

depending upon the circumstances. Nothing was simple, nothing straightforward." (*In the Lake of the Moon*, p. 174)

Haydon's Latin American sojourns offer images and perspectives of an Anglo-American participant observer of Latin America that underscore the ambiguous nature of our southern neighbors. ■■■■■■■■■

The American Midwest of P. M. Carlson

"What made you look there?"
"The geography just hit me," [Marty] explained.
(Bloodstream, p. 202)

ARTY (MARTINE LAFORTE) HOPKINS, deputy sheriff of Nichols County, Indiana, is a second female cop. To date, the series consists of only two books—*Gravestone* and *Bloodstream*. P. M. Carlson does a fine job sharing her images of small-town, middle America, in southern Indiana, through the eyes of Deputy Sheriff Marty Hopkins. Beautiful countryside, pocketed with caves and abandoned quarries, serves as an essential element in the plot of the first mystery, while the White River serves the same purpose for the second. One is constantly reminded of scenes from *Breaking Away*, which is alluded to in the first novel. Just as that movie caught the spirit of middle America, so do the murder mysteries in this series, with a heavy reliance on physical geography.

Southern Indiana is "limestone country," and Carlson puts the geography and the geology of the region to good use. The landscape provides an integral part of the plot. Caves in *Gravestone* provide the grave for a body long missing, the hideout for the main villain, and the history of the cave as a play area for kids years ago helps reveal the identity of the murderer.

The important role of the landscape also allows a brief lesson about the importance of these limestone deposits and their evolution, as Professor Wolfe explains to Marty:

> "Do you know about limestone, Martine?"
> "Some. My grandfather worked in the quarries. Dug out the Empire State Building, he always said."
> "You know it's ancient sea bottom."
> "Imagine a great sea over America, Pennsylvania to Nebraska. Suppose you sail a boat out to the middle and drop anchor over Indiana . . . There are lots of little shelled sea creatures below you . . . For five million years it'll be serene. The animals grow, make shells from dissolved minerals, reproduce, die. The waters grind and smooth their shells and cement them into stone. Generation after generation. Then underground forces push up the seabed to form land. And here we are." (*Gravestone,* pp. 34–35)

The transition from southern Indiana to Kentucky is nicely captured on Marty's trips to inquire about an apparent kidnapping that seems to be related to her cases in Indiana.

> When you cross the Ohio River south into Kentucky, the landscape doesn't change a lot. It's still limestone country, but the bands of stone below the russet soil are not as thick and flawless as the fine-grained belt that surfaces in southern Indiana . . . But the rains fall in Kentucky too, and the water sculpts the limestone as it does in Indiana, carving rounded hills and valleys and sinkholes, tunneling out caves as magnificent as Mammoth, and decomposing the top layer of rock into soil suitable for growing a pelt of maples and redbuds and wild grapevines . . . in southern Indiana, as in Kentucky, the soil sprouts mint for juleps and makes the bluegrass green. (*Gravestone,* p. 129)

Whereas the limestone caves are essential to the story line in *Gravestone,* the White River (a real river in fictitious Nichols County) becomes a critical part of all phases of the crime in *Bloodstream.* Carlson uses the river to avail readers of a number of lessons on the physical geography of the region.

> Like most rivers in Indiana, the East Fork of the White River begins quietly in the glacier-smoothed eastern uplands that drain into small, slow creeks and rivulets. The streamlets crawl south and west, joining other gurgly young creeks, turning frisky as they hit the rough, eroded limestone country of southern Indi-

Although different in approach and perspective, both of these series share crimes and resolutions steeped with the culture and geography of New Orleans and southern Louisiana. If Robicheaux is constantly involved with pimps, prostitutes, drug dealers, and other assorted low-lifes, Langdon is usually entrenched with the wealthy, the powerful, and the prominent. Together, the two series complement each other nicely.

The Houston and Latin America of David Lindsey

Mexico City had become a grotesquerie.
And yet, as if still clinging to the dualism that
pervaded the ancient Aztec theology that had flourished
in the once-beautiful Valley of Mexico
six hundred years earlier, the city had another face.
(In the Lake of the Moon, P. 164)

FOUR HUNDRED MILES to the west of New Orleans, David Lindsey gives us Homicide Detective Stuart Haydon of the Houston Police Department. Fortunately for the reader, Stuart often gets involved in cases that take him out of Houston to Latin America. It requires a stretch of the imagination to define these novels as police procedurals because most of the plot occurs out of Stuart's jurisdiction. But the sense of place is so poignant, and Latin America is sufficiently underrepresented in the police procedural literature, that the Haydon series adds a welcome component to our regional ramblings.

First, however, we learn of Houston itself—often quick glimpses from time to time. In setting the weather and mood of the city in fall, Haydon notes that Houston

> was beginning to smell fusty, and the trunks of the trees were sprouting a fine, linty pelt of Kelly-green lichen. Every color exposed to the weather was two shades darker from the incessant

> moisture . . . The bayous had risen and stayed up, and some parts of the city flooded every few days. (*In the Lake of the Moon,* p. 6)

A favorite Lindsey subplot is the contrast between the rich and the poor, employing images of neighborhoods to bring home the contrast. It also provides an opportunity to share displeasure at the development of downtown Houston over the years.

> Haydon's route home did not require him to travel the freeways that embraced Houston like tentacles of sargasso reaching in from the Gulf of Mexico to strangle the city with congested veins of simmering cement. Instead he turned from the blistering downtown canyons of steel and glass. (*A Cold Mind,* p. 47)

The tour of Houston takes us to some pretty exclusive areas, such as the St. Remy health spa.

> Haydon stopped at the guardhouse, identified himself, and pulled into the narrow lane leading through the towering pinewoods that formed an arboretum over the entire estate. For a while the lane followed alongside the jogging trail, which was covered with Superturf and twisted through the forest like a mossy carpet. Occasionally the pines would open to a meadow, with cleanly mown emerald grass. (*A Cold Mind,* p. 137)

A short while later, Haydon finds himself in a very different Houston: "He squinted back in the glaring sun at the Marsdon Court housing projects. Their cement-block dormitory-style architecture set the tone of hopelessness that smothered the barrio neighborhood like a stultifying tropical heat wave" (*A Cold Mind,* p. 146).

Much of the focus of this series centers geographically on Latin America—Mexico City in *In the Lake of the Moon* and Guatemala in *Body of Truth.* Quite different senses of place emerge from each of these novels. Not coincidentally, his ramblings outside of the United States take place after the death of his partner, a time when not only is he on his own, and thus capable of these unusual detours out-of-country, but also a time when he goes through a great deal of self-examination, where physical journey parallels psychological journey.

The Mexico City of *In the Lake of the Moon* is presented ambiguously, offered as a love-hate relationship between Haydon and the city.

Lindsey intertwines outright description, mood, climate, history, mythology and folklore—quite detailed and quite evocative: "Haydon had a lifelong knowledge of this sprawling, out-of-control metropolis and an alternating affection and abhorrence for all that it represented" (*In the Lake of the Moon*, p. 163).

One can sense this "dualism" in the manner in which Haydon captures the layout of the city.

> The city's layout owed more to romance, emotion, and caprice than it did to logic. . . . It was a nightmare system . . . that raised the metaphor of the city as labyrinth to a higher level of meaning. Sometimes it bordered on the abstract. (*In the Lake of the Moon*, pp. 166–167)

The Guatemala of *Body of Truth* does not receive so sympathetic an interpretation. The sense of place portrayed in the novel is primarily political, combining a sense of terrorism and repression with a touch of American duplicity in an unambiguously negative perspective.

One of Haydon's contacts in Guatemala, an ex-CIA officer, explains the situation quite simply.

> "One: the army controls everything . . . Two: there's a tiny wealthy elite that holds most of the arable land and largely controls the economic fortunes of the country. Three: the military upholds the inequitable situation, and between the military and this mercantile elite there is an easy romance . . . Four: the big losers are 90 percent of the population. Five: none of this is about to change. The civilian 'government' is a real farce." (*Body of Truth*, p. 117)

As adjunct to the overall political situation in Guatemala, we are reminded throughout the novel that torture and killing are part of the daily routine, and that impending visits by government-supported death squads are a way of life for many Guatemalans. We are also reminded of the extreme poverty and despicable living conditions of most Guatemalans. The support of the U.S. government in the name of trying to bring democracy to Guatemala cannot be avoided. The final resolution of the deaths of a number of people are accepted as expected within this context.

Detailed character development is one of the trademarks of the

Haydon series and one of its great appeals. In *In the Lake of the Moon,* the murder directly involves Stuart Haydon and the history of his family. As a result, we learn much about Stuart Haydon and his father, a bit about his wife, Nina, and his maid Gabriella, and much less about his mother and her side of the family. Although Haydon is independently wealthy, the reader is not put off by it because Haydon is a real person going through a quest for the truth about his father's past, revealing a side of his father he knew nothing about. The story line is universal; it could happen to you!

Even when the story line is not directly about Haydon and his family, character development is quite detailed. The actual murders don't take place until late in the novels, and by then we're pretty sure who's going to be murdered. The only questions are how, why, and who else. In all cases, the character development is used effectively by Lindsey to provide us with a history of the places involved.

In one instance, Lindsey relates the history of Mexico by focusing on the history of one of its streets.

> Licenciado Verdad was a short street with a long history. In the early sixteenth century when Hernán Cortés with his Spanish soldiers and Catholic priests crossed the long causeway and entered the Aztec capital of Tenochtitlan. . . .
>
> After the Spaniards conquered the Aztec empire in 1521, they razed the ancient city . . . and then modern Mexico replaced the colony, but the geographic center of each of these empires remained the same, the heart of Mexico City. The narrow, alley-like street of Licenciado Verdad had seen it all. (*In the Lake of the Moon,* p. 301)

Although native folklore and mythology do not play the role they do in Hillerman or Hager, there is a heavy dose of native Central American culture, used basically to help explain the nature of Mexican society today.

> "So, what did the Indians learn about life, about themselves? At the very core of the later Mesoamerican religions is the concept of duality . . . the unity of basically opposed principles. Life and death, light and darkness, good and evil, up and down, yes and no. Each god was more than one god; sometimes he was many,

depending upon the circumstances. Nothing was simple, nothing straightforward." (*In the Lake of the Moon,* p. 174)

Haydon's Latin American sojourns offer images and perspectives of an Anglo-American participant observer of Latin America that underscore the ambiguous nature of our southern neighbors. ■■■■■■■■■■

The American Midwest of P. M. Carlson

"What made you look there?"
"The geography just hit me," [Marty] explained.
(Bloodstream, P. 202)

ARTY (MARTINE LAFORTE) HOPKINS, deputy sheriff of Nichols County, Indiana, is a second female cop. To date, the series consists of only two books—*Gravestone* and *Bloodstream.* P. M. Carlson does a fine job sharing her images of small-town, middle America, in southern Indiana, through the eyes of Deputy Sheriff Marty Hopkins. Beautiful countryside, pocketed with caves and abandoned quarries, serves as an essential element in the plot of the first mystery, while the White River serves the same purpose for the second. One is constantly reminded of scenes from *Breaking Away,* which is alluded to in the first novel. Just as that movie caught the spirit of middle America, so do the murder mysteries in this series, with a heavy reliance on physical geography.

Southern Indiana is "limestone country," and Carlson puts the geography and the geology of the region to good use. The landscape provides an integral part of the plot. Caves in *Gravestone* provide the grave for a body long missing, the hideout for the main villain, and the history of the cave as a play area for kids years ago helps reveal the identity of the murderer.

The important role of the landscape also allows a brief lesson about the importance of these limestone deposits and their evolution, as Professor Wolfe explains to Marty:

> "Do you know about limestone, Martine?"
>
> "Some. My grandfather worked in the quarries. Dug out the Empire State Building, he always said."
>
> "You know it's ancient sea bottom."
>
> "Imagine a great sea over America, Pennsylvania to Nebraska. Suppose you sail a boat out to the middle and drop anchor over Indiana . . . There are lots of little shelled sea creatures below you . . . For five million years it'll be serene. The animals grow, make shells from dissolved minerals, reproduce, die. The waters grind and smooth their shells and cement them into stone. Generation after generation. Then underground forces push up the seabed to form land. And here we are." (*Gravestone,* pp. 34–35)

The transition from southern Indiana to Kentucky is nicely captured on Marty's trips to inquire about an apparent kidnapping that seems to be related to her cases in Indiana.

> When you cross the Ohio River south into Kentucky, the landscape doesn't change a lot. It's still limestone country, but the bands of stone below the russet soil are not as thick and flawless as the fine-grained belt that surfaces in southern Indiana . . . But the rains fall in Kentucky too, and the water sculpts the limestone as it does in Indiana, carving rounded hills and valleys and sinkholes, tunneling out caves as magnificent as Mammoth, and decomposing the top layer of rock into soil suitable for growing a pelt of maples and redbuds and wild grapevines . . . in southern Indiana, as in Kentucky, the soil sprouts mint for juleps and makes the bluegrass green. (*Gravestone,* p. 129)

Whereas the limestone caves are essential to the story line in *Gravestone,* the White River (a real river in fictitious Nichols County) becomes a critical part of all phases of the crime in *Bloodstream.* Carlson uses the river to avail readers of a number of lessons on the physical geography of the region.

> Like most rivers in Indiana, the East Fork of the White River begins quietly in the glacier-smoothed eastern uplands that drain into small, slow creeks and rivulets. The streamlets crawl south and west, joining other gurgly young creeks, turning frisky as they hit the rough, eroded limestone country of southern Indi-

> ana. They . . . [bounce] into Nichols County as a full-fledged river . . . and [joins] the West Fork to head out for the Wabash, the Ohio, the Mississippi, and the sea. (*Bloodstream*, p. 3)

Yet for Carlson these are more than simply straightforward lessons in physical geography; rather, the river becomes a character in the plot, is personified as one who helps to solve the mystery: "the river was chewing away . . . it had uncovered a coffin, rotted it open, stolen the bones" (*Bloodstream*, p. 312).

Like the story line in Susan Dunlap's Jill Smith series, the story line of the murder investigation is intertwined with Marty's struggle as a female cop, attempting to succeed in a male-dominated environment. One complication for Marty is the multiple roles she tries to balance. She is not only working to succeed as a homicide detective, she is also trying to succeed in the traditional roles of wife and mother for a self-centered, unreliable husband and a confused daughter. This very personal side story helps to develop the main character in such a way that it endears her to the reader. Although much of the friction at home can be attributed to the self-centeredness of the husband and the age of the daughter, Marty sees herself failing as wife and mother. Professor Wolfe's admonition to her to "rescue herself" not only applies to the murder case but to her personal life as well.

As if struggling to raise a teenage daughter alone isn't complicated enough, Carlson throws another twist into Marty's personal life in the second novel, when her rambling husband, Brad, finally finds himself and his life's dream—a job as a disc jockey near Memphis. He wants Marty to give up her job in Nichols County and join him with their daughter. The conflict interplays throughout the novel since being a deputy sheriff has become a career for Marty, and the debate ensues over why one career is more important than the other, with the additional pressure to move coming from her daughter so that the family can be together again.

In *Gravestone,* the dominant social theme that plays a vital role in the plot is the legacy of the Ku Klux Klan. This is southern Indiana, where the Klan has always been strong. Although the Klan was discredited during the 1960s and 70s, the story line reminds us that deeply felt support lingers if not for the Klan itself and its more despicable practices, then at least for some of its principles. For some residents, the values of the Klan represent a return to a more moral and pristine past. For others, these values are simply a cop out for the failures of white males.

One of the main thrusts of this part of the plot is the transition that has occurred, where most law-abiding citizens abhor the Klan and what it stands for.

Sheriff Cochran exemplifies this new kind of law and order, credited with almost single-handedly putting the Klan out of business in Nichols County fifteen years earlier. Yet feelings of the need to protect racial purity, especially against blacks and Jews, still remain. The presence of a mixed marriage—a Jewish man to a black woman—resulting in the murder of the man, precipitates the story. Two seemingly unrelated cases merge into one by the end of the novel, tying together incest, revenge, suicide, murder, the Klan, and an intense rescue scene to bring the mystery to conclusion.

Here again, the characters represent various aspects of the changing nature of small-town middle America. The more liberal-minded sheriff, the wealthy, but dying remnant of the old, Klan-based south, the professor who has turned her back on society for the most part, and our heroine who struggles to rescue herself as she helps define new roles for women in previously male-dominated careers like law enforcement.

In *Bloodstream,* the ramifications of extreme child abuse and religious fanaticism take over as dominant themes. Here again, Marty's dual roles as mother and cop provide a different kind of perspective as she tussles with issues of the safety of her daughter and the value of keeping the family together. In many ways, these are not just women's issues, but issues of importance to all parents, helping to remove any objective perspective on the part of the reader. For many, it becomes personal as well. ■■■■■■■■■■■■■■■■■■■■■■■■■■■■■■■■■■■■■

The American Northwest of J. A. Jance

The State of Washington is divided into two parts, east of the mountains and west of the mountains. They could just as well be separate countries.
(Injustice for All, P. 89)

IF ANYONE FITS the description of writing a bits-and-pieces approach to place-based novels, it's J. A. Jance. Upon finishing one of her police procedurals, the reader's immediate reaction is that the cultural context is weak. Yet you have a feeling that you have been there, in the rain and drizzle, out among the lakes and islands and bridges, traveling on floats and ferries—maybe the cultural context isn't all that weak after all. In reality, it is not. J. A. Jance is a mistress of the subtle, almost unnoticeable feeding of short snippets into the reader's subconscious as he or she concentrates on fairly strong character development and intricate plot twists along the way to solving the murder.

Jance helps to shatter our images of a serene and tranquil American Northwest with the help of Seattle Homicide Detective J. P. Beaumont—Beau to his friends and readers of the series. While her approach to sense of place is subtle, Jance's approach to character development isn't. Much of the allure of the series comes from the strong development of characters—not just the hero, and not all of whom live throughout the series. This is clearly a series in which the reader feels he or she is part of the extended Beaumont family.

Although her approach to sense of place and character development has been consistent throughout the series, there has been a change in the length of her novels. The first ten were approximately two hundred pages, like the novels of Kaminsky, Upfield, and other shorter-length novelists. In *Failure to Appear,* she moves into the range of three hundred plus pages, more in keeping with Hillerman, Lindsey, and Smith. This allows her a little more room for character and plot development, and it certainly does nothing to detract from keeping reader interest throughout the novel.

In the J. P. Beaumont series, we find the classic strengths of the police procedural genre—real-life police procedures, teamwork as the basis for the resolution of the crime, realistic descriptions of the murders, and strong character development. For the place-based procedural, we find the subtle interweaving of a sense of Seattle and its environs. With a spatial eye for patterns, Jance uses physical geography to divide Washington into two cultural realms.

> West of the mountains is a fast-track megalopolis that is gradually encroaching on every inch of open space. East of the mountains seems like a chunk of the Midwest that has been transported and reassembled between the Cascades and the Rockies. (*Injustice for All,* p. 89)

J. P. Beaumont has come into money as a result of the tragic death of his second wife, Ann Corley, a loss that haunts him through many of the following novels. This wealth is not an alienating characteristic, however. Beau himself often feels uneasy about his wealth, which helps us relate to him on a personal level. Beau comes across as a hard-boiled detective type, trying to adjust to the new male image of the 1980s. He certainly has personal problems that command a good deal of our reading time and often play into the detection of the murderer. His first wife, Karen, and his children grow up before our eyes during the series, with finally a wedding and a granddaughter in *Failure to Appear,* if not in that order. A good bit of the book deals with a father's coming to grips with alternative lifestyles and the growing up of his daughter Kelly.

As an extended member of the Beaumont and Peters families, Ralph Ames, Beau's lawyer and friend, is an integral part of resolving many of the family issues, such as securing custody of the girls, defending accused friends, and even helping to solve some of the murders. The close emotional bond and friendship between Beau and Ames arises because Ralph was originally Ann Corley's lawyer, and he stayed on to help Beau after her death. Characters appear throughout the series, which gives Jance the opportunity to neatly tie the knot on the cord that binds the readers to the main characters. Jance has done a marvelous job at making these characters more than just characters in a novel. They are our friends, and we look forward to finding out what has happened to them since last we met.

As for Seattle and its environs, and the sense of place created throughout the series, Jance makes good use of those characteristics that define the region—hills and mountains, drizzle and rain, lush green vegetation, and the numerous bays, inlets, sounds, and rivers. As in any good place-based police procedural, these characteristics are more than just backdrops. They are often an integral part of the commission, investigation, and resolution of the crime.

On a number of occasions, the series departs Seattle proper—to the suburbs, to small towns and resorts in Washington, even to Ashland, Oregon, where the sense of place is provided by the world-famous Shakespearean Festival.

> The tourist guidebooks all say that Ashland is a lovely, picturesque place. Quaint, I believe, is the operative word. The shady tree-lined streets showcase prosperous looking, newly rehabilitated but authentically Victorian houses of the gingerbread variety. Most of the bigger ones seem to have been converted into bed-and-breakfast establishments. (*Failure to Appear*, p. 17)

J. A. Jance is not resistant to injecting a number of social issues into the plot. Because the series is carried out in the first person, through the words and thoughts of Beau, issues are generally provided as he would deal with them personally. Alcohol, drugs, gays, and other kinds of alternative lifestyles become subplots throughout the series. His own alcoholism and the alcoholism of others close to him, drug use by victims and suspects, and alternative lifestyles of victims, suspects, and family members reveal the trials and tribulations of a middle-aged white male as he deals with the changing mores of society.

Several devices used for heightening reader interest are quite effective. In several of the novels, the murders involve someone close to Beau. His daughter's friend is a murder suspect in *Failure to Appear*. His lover is murdered in *Injustice for All*, and another lover is a murder suspect in *Taking the Fifth*. Jance's portrayal of Beau is a bit of a throwback to the hard-boiled detective genre. He's a hard drinker, and although he's often hungover, it doesn't seem to affect him—until at some point he realizes he's an alcoholic. He often gets into bed with beautiful women involved in the murder. Jance's use of contrasts for effect, especially between high-rent and low-rent districts in Seattle and between urban Seattle and its rural environs, really helps to emblazon a sense of place.

J. A. Jance's J. P. Beaumont has provided a very popular series that clearly counts as a place-based police procedural. It also has provided a sense of Seattle and the American Northwest to tens of thousands of readers. ■■

Susan Dunlap's Berkeley

If you can't enjoy the peculiarities of your fellow citizens, you'd better not live in Berkeley. Certainly you should not be a police officer there.
(Sudden Exposure, P. 1)

ALTHOUGH SEPARATED FROM San Francisco by only the Bay Bridge and a couple of off ramps on Interstate 80, Berkeley is another world. For millions of people, images of the 1960s still define the character of this Bay Area enclave. And although remnants of the 1960s can be found pestering locals and tourists alike along Telegraph Avenue, facets of this intriguing center of one of the nation's top universities are relatively unknown to those who have never spent time there. Susan Dunlap's writing goes a long way toward filling in the blanks of our sense of a particularly unusual, intriguing, and ambiguous place.

Dunlap presents us with our third female homicide detective, Jill Smith, whose career we follow from the very beginning. The series begins just as Jill has been promoted from traffic to homicide. Later, she will return to being a beat cop. Although the difficulties of being a woman in a predominantly man's world are dealt with throughout the novels, they never displace the resolution of the crime as the main focus of the story line.

In a way, Jill's character represents the antagonisms of the Berkeley community. Many of the stereotypes about Berkeley and its inhabitants are only partly true. Smith exemplifies how police procedurals are often used to break the stereotypes. The reality is far more complex. Jill, for example, is divorced and drives a vintage VW bug, which we would expect; yet she is a junk food addict, which breaks down our traditional image of the average Berkeleyite.

She often displays respect for certain aspects of the typical Berkeley lifestyle:

> There is a typical Berkeley matron like Rue Driscoll—not about to dye her hair or dabble in plastic surgery, but having no intention of growing old with quiet grace either. She can be seen striding along Shattuck or College Avenue in Birkenstocks, on her way to a meeting to preserve public access to San Francisco Bay,

> to fight Medicare cuts, or to plan another protest at the Diablo Canyon nuclear plant . . . "Little old lady" is something she has no intention of becoming. (*A Dinner To Die For*, p. 42)

Yet it is just as easy to find her poking fun at the northern California lifestyle:

> [H]undreds of Northern Californians had changed their names in the last decade. Animal rights advocates called themselves Laughing Otter, the ecology-minded became Singing Rainbow or Green Meadow, and those who found gurus switched from Jim, Jane, and Jerry to Ananda, Jyoti, and Ram. (*A Dinner To Die For*, p. 153)

Dunlap gives us a glimpse of a Berkeley in transition from its well-known past to its present image, providing descriptions of the different Berkeleys that emerged by the 1980s, even going so far as to comment on the changing nature of that most famous of all Berkeley streets, Telegraph Avenue.

> The sun was bright, the air still post-storm fresh. Telegraph Avenue looked just cleaned. The rain had washed the pizza papers and dog turds off the sidewalk. Many people complained that the avenue they knew was disappearing in a wave of gentrification. New boutiques and croissant shops boasted bright tile, clean windows, and orange neon signs. They gave the impression that the entire avenue could be hosed clean. (*Not Exactly a Brahmin*, p. 116)

Not only does she use landscape and neighborhood descriptions to portray the multifaceted personality of present-day Berkeley, she also uses characters as representatives of the different social groups that one finds in this complex suburb of San Francisco, whose presence can never be totally ignored. From time to time, we get vivid descriptions of the magnificent views of Baghdad by the Bay as seen by many a Berkeleyite: "San Francisco Bay glistened. The Bay Bridge arched like a silver filagree necklace, dropping down onto the kelly green of Treasure Island" (*Not Exactly a Brahmin*, p. 88).

Yet Dunlap does not resort to relying on San Francisco for locale. This is first and foremost a series that takes place in Berkeley, and the eccentricities and peculiarities of the place are important pieces of the overall story line. Other real-life situations are also integral to the plot—the weather, especially the famous Bay Area fog, and the changing nature of Berkeley society (cutbacks in the city budget as it impacts the police department, for example). ■■■■■■■■■■■■■■■■■■■■■■■■■■■■■

The Canadian North of Scott Young

The newcomers, including many from southern Canada, go north for all kinds of reasons, from leaving trouble behind to looking for a new meaning in life.
(The Shaman's Knife, P. 51)

THROUGH THE EXPLOITS of Royal Canadian Mounted Police (RCMP) Inspector Matteesie Kitologitak, Scott Young brings to the police procedural many of the qualities that have made Hillerman so popular with his depictions of Navajo culture in the American Southwest. Like Hillerman, Young provides striking portraits of the Canadian North, a sensitive, yet realistic interpretation of Inuit culture, and a subplot that examines the environmental and cultural impact of European modernization on the Canadian North. Whereas Hillerman uses two characters to portray the conflict between the traditional and the modern in Navajo culture, Young embodies this antagonism in a single hero—Matteesie Kitologitak.

Several characteristics of the Canadian North come through loud and clear in the writing of Young—it is a vast and inhospitable place, where civilization is separated by great distances. The map produced in the first few pages gives us our first taste—Young leaves in an outline of the United States as a point of reference for the distances involved in traveling between the places mentioned in the book, and the Arctic Circle alerts us of just how far north we are. Quickly we return to our introductory geography texts to remind ourselves just exactly what this means. Likewise, the harsh environment of the Canadian North is very fragile, and the changes wrought by modernization can have devastating effects. The same may be said for the Native cultures as well. There are strong parallels between Native habitats and Native cultures.

Native Canadian Indian culture, particularly Inuit culture, is essential to solving the case. Through Matteesie, Young provides a skeptical, yet reverent treatment of traditions and myths, bound up with glimpses back into the history of the white suppression of Native Americans, whose spiritual leaders were the

> men and women who, as shamans, were the chief instruments of the fairly complicated set of tribal beliefs and legends that all our people once lived by. That was before Anglican and Roman Catholic missionaries spread through the Arctic building churches and teaching the Christian version of what life was all about.
>
> But whatever the missionaries did, or thought they did . . . well, you can't wipe out ten or twenty or thirty centuries of beliefs. . . . There were shamans still around. (*The Shaman's Knife,* p. 21)

The resurgence of traditional customs and practices is a secondary part of this subplot, something to reunite Native American peoples after centuries of subjugation. Even the acculturated Matteesie uses the symbolism of the fisherman at the seal hole to represent his relationship to the murder investigation: "I was a native, too, sitting at a seal hole. Make the slightest false move, and no seal, or in my case no answer. I had all day" (*The Shaman's Knife,* p. 180).

The major disappointment with this series is the lack of availability. *The Shaman's Knife* can be found in many bookstores. The other two novels in the series, unfortunately, are out of print. There is no indication whether the series will continue so that we might gain a more complete picture of Matteesie and the places he frequents. ■ ■ ■ ■ ■ ■ ■

The Mexico of Paco Taibo II

[T]he important thing is not the crimes, but (as in every Mexican crime novel) the context.
(Life Itself, P. 144)

THE MURDER MYSTERY FAN who is searching for a different kind of approach to the police procedural will be rewarded by traveling south of the border to the Mexico of Paco Ignacio Taibo II. Although he does not have a series based on a single hero, the unusual style and story lines provide a nontraditional interpretation by one of the few authors from the lesser-developed world who has found an audience in the developed world. Enjoying immense popularity in Latin America, several of Taibo's works have been translated into English. Examples from some of his works in print in the United States serve to underscore the very different context for his contributions to the murder mystery genre. It may be fair to say that context is the key to the writings of Taibo. With this in mind, the murder mysteries of Paco Taibo add a unique and intriguing glimpse into crime as perceived in Mexico and as interpreted by a native Mexican—a very different approach from David Lindsey, for example. His most famous character, Hector Shayne, is a private eye. But Taibo does dabble in the procedural genre in a unique way.

In *Life Itself,* Taibo introduces Jose Daniel Fierro, a mystery writer who is asked by the citizens of Santa Ana to serve as sheriff, primarily to help explain to the outside world what they are trying to do with the installation of a radical local government. Although Santa Ana is a fictional city, it is representative of northern Mexico's challenge to the incumbent PRI, which in actual fact has held power over Mexico for most of this century; the murders take a backseat to the context, in this case the context of a radical challenge to PRI control. Taibo provides a sympathetic treatment of the situation in Santa Ana; although clearly a radical experiment in local government, the novel portrays it basically as an attempt by the locals to control their own destiny against the crime and corruption of the central government. In *Life Itself* there is little landscape description; rather, the narrative description provides primarily a political sense of place. In a number of places, the description of the urban landscape serves as a reminder of the political turmoil that has led to this most unusual scenario.

> The whole town was covered with graffiti: sidewalks, fences, light poles, columns, even some low rooftops, house fronts, walls, trees. All painted over many times and by many hands, in many styles; differing hands piling up slogans and signs, advice and insult, calls to conscience, logos of the Popular Organization, calls to the future, memories of the past. (*Life Itself,* p. 13)

Initially, it is hard to know whether or not to take the novel seriously, especially given the bizarre setup and the background of our hero, the mystery writer turned "policeman." At times the novel is quite humorous, seemingly satirical, but gets the point across; and in the end, it paints a clearer picture of the relationship between some regions of Mexico and the central government, a picture that is particularly sympathetic to the locals' cause.

One fairly effective means for narrative description is Taibo's use of notes and personal letters. He intermingles the day-to-day events in Santa Ana with his "notes for the history of the radical city government of Santa Ana" and with letters to his wife, Ana, back in Mexico City. In one such letter, written upon his arrival in Santa Ana, Jose Daniel underscores his approach to being the new sheriff of this renegade community: "I regard myself as a joke, but I take very seriously the experiment of the radical city government of Santa Ana" (*Life Itself,* p. 26).

This is one case in which the resolution of the murder takes a backseat to the context for murder, a commentary on the wretched conditions in the Mexican hinterland and the devastating effects of PRI control throughout Mexico. It is a story well worth the read. ■■■■■■

Conclusion

These ten series capture the varied nature of this American born-and-bred genre of detective fiction. Several defining characteristics set each of them apart as police procedurals—police officers employing real-life police procedures to solve murders. Yet there is as much, if not more, variability between series, and often within the same series as well. This variability is a reflection of the authors' innovation and of the fertile ground of the human condition, as well as the complexity of different places for murder. The evolution of the genre has seen a series of responses to challenges made to the convention. What happens if the detective is a woman? What happens if the detective's family is personally involved in the murder? What happens if a detective's personal life conflicts with his or her professional life? And so on. The response that each challenge elicits results in a new variant of the genre, while maintaining a few basic and common components.

At the same time, place acts as an important component in all of these series. The physical and cultural geographies of the settings for murder incorporate aspects of landforms and geomorphology, cultural and historical characteristics, with emphasis on ethnic, religious, and political traits. And the importance of searching for patterns, looking for spatial relationships, and using maps are often infused as key tools in the resolution of these crimes.

Like many aspects of American culture, the police procedural has been exported to other parts of the world as well as to other periods in history. Sometimes this has been accomplished by American authors but more often by native authors. In either case, foreign adaptations of the police procedural provide examples of even more variety in the genre. They also increase the necessity for authentic sense of place.

Murder in the United Kingdom and Ireland
Turnbull
Robinson
Gill
Dexter
James
Chapter 4

Although the origins of the murder mystery and the police procedural genres are rooted in the United States, most mystery fans associate the solving of literary murders with Britain, where Arthur Conan Doyle, Agatha Christie, Dorothy Sayers, and others popularized this form of literature. The police procedural format was imported to the British Isles in the 1950s, and contemporary British authors adapted to this new twist in the murder mystery, carrying on the tradition of popular literary murder emanating from Britain. At the same time, many of these authors have become effective practitioners of sense of place. The selection includes five authors who take us to London, Oxford, rural England, Scotland, and Ireland. We begin in London with the writing of P. D. James, clearly one of Britain's most popular contemporary crime story writers. A very different urban place is Oxford, the setting for murders plotted by Colin Dexter. In contrast to the vivid sense of place in James' London and Dexter's Oxford, rural England is portrayed in the writings of Peter Robinson, who takes us to the Yorkshire countryside. Robinson's sense of place is offered as fragments, niggardly spliced into the fabric of the plot; yet a number of characteristics are shared by the rural locales. These settings are pictured as quiet, even serene, with a much slower pace of life. Rural authors often contrast this serenity with London. The intrusion of urban life into the countryside is a recurring subplot. The murder is also portrayed as an intrusion into the serenity of country life. In all three of these, the influence of the more traditional British locked-door murder mystery, reminiscent of Doyle and Christie, is evident. Particularly British is the need in all of these novels to explain the resolution in intricate detail. Yet the dependence on procedures dictates that these do fit into the place-based police procedural genre. Outside of England, we travel to Scotland and Ireland. Peter Turnbull bases his procedural series in Glasgow, while the murders of Bartholomew Gill are set in and around Dublin. Outside of England, nationalism, both Scottish and Irish, becomes an important subplot.

In contrast to the tendency of female American writers to create female lead characters, all lead characters in these series are male, even the P. D. James' series, well in keeping with the British tradition of Christie and Sayers. Also, environmental issues are missing to a large extent, except to the degree that they are included within the subplot of urban intrusion into the serenity of the British countryside. ■ ■ ■ ■ ■ ■

The London of P. D. James

For those who sought symbols in inanimate objects,
its message was both simple and expedient,
that man, by his own intelligence and his own efforts,
could understand and master his world.
(Devices and Desires, P. 63)

P. D. JAMES, OFTEN referred to as the "Queen of Crime," is known first and foremost for her intricate plots and detailed character development. The Adam Dalgliesh series spans over thirty years, during which time it has undergone several major changes. Whereas the earlier mysteries of the series tended to range in the mid-200-page category, her more recent works double that. In paperback, *Devices and Desires* approaches 470 pages, while *Original Sin* exceeds 540 pages. All of her novels, no matter what the length, stand as excellent examples of James' ability to provide intricate plots and detailed characters. We are given careful and elaborate histories of all possible suspects and investigators, numbering over a dozen in each case, contributing to the length of these novels.

In this regard, James differs from her British tradition in which characters were kept as superficial as possible, revealing only what was absolutely necessary. James is a bridge between the old and the new, mixing certain aspects of the traditional British whodunit with the new demands of the police procedural. Nowhere is this blending more evident than in the person of the main protagonist.

The lead character of the series is Adam Dalgliesh, Criminal Investigations Division (CID), Scotland Yard, who works his way from commander to superintendent over the years. In Dalgliesh, James blends aspects of the Great Detective genre with qualities of the police procedural. By the beginning of the series, Dalgliesh is already a widower, his wife and son having died in childbirth. This probably explains his reticence toward children some years later: "He was not a man who liked children and he found the company of most of them insupportable after a very brief time" (*A Mind to Murder,* p. 154). Although hints of possible romantic trysts arise from time to time, they remain only possibilities, and romantic involvement for Dalgliesh does not become a subplot of the series.

What does become a recurrent and underlying theme of the series is the contradictory nature of the Dalgliesh character. On the one hand, reminiscent of Sherlock Holmes, he is an artist—a published poet of some reputation. On the other hand, his deductive skills are based on good, solid, and methodical police work, although there are hints of the Great Detective characters as well: "Dalgliesh pointed out mildly that it was also common sense, the basis of all sound police work" (*A Mind to Murder,* p. 81).

This might be translated into "Elementary, my dear Watson." These hints at the powers of deductive reasoning notwithstanding, the key to Dalgliesh's success as a policeman comes from sound, fundamental police procedures: "And so it went on; the patient questioning; the meticulous taking of notes, the close watch of suspects' eyes and hands for the revealing flicker of fear, the tensed reaction to an unwelcome change of emphasis" (*A Mind to Murder,* p. 95).

Dalgliesh is clearly in the procedural mode even with hints of the Great Detective from time to time. Also, reminiscent of the Great Detective approach is the nature of the relationship between the main protagonist and his assistant, again reminiscent of the relationship between Holmes and Watson. Traditionally, the assistant served primarily as a dupe for the brilliance of the Great Detective. James is quick to point out that the basic nature of the relationship has changed, as epitomized by the relationship between Dalgliesh and his assistant, Martin.

> Watching [Dalgliesh and Martin] at work a casual observer might have been misled into the facile assumption that Martin was merely a foil for the younger, more successful man. Those at the Yard who knew them both judged differently . . . No one watch-

> ing Dalgliesh at work could fail to recognize his intelligence. With Martin one was less sure . . . But he had qualities that made him an admirable detective . . . It was Martin's job to help solve [murders] and, patiently and uncensoriously, that is what he did. (*A Mind to Murder,* pp. 101–102)

Although Dalgliesh will change assistants later in the series, the nature of the working relationship of the team is set. Dalgliesh is clearly the lead investigator, possessing intelligence and powers of deduction. His assistants are, however, valuable and contributing members of the team effort to solve cases. Eventually, a female officer will rise to the stature of assistant to Dalgliesh.

One interesting twist in P. D. James is that the focus is not solely on the investigator. The reader gets the story from the perspective of many of the characters, often even from the perspective of the victim. In the early pages of *Devices and Desires,* we get the lead into one of the murders from the victim herself.

If James is a mistress of personifying the blending of traditional British with police procedural characteristics, she is just as much a mistress of sense of place. As with most authors of police procedurals, James uses outright narrative description to integrate a bit of history. Nothing in London suggests more about history than the Thames: "From her earliest years her father had told her stories of the [Thames], which for him had been almost an obsession, a great artery endlessly fascinating, constantly changing, bearing on its strong tide the whole history of England" (*Original Sin,* p. 91).

Not only do we get a sense of the Thames and various venues around present-day London, we also find an interesting twist in our sense of place, when the characters in *Original Sin* all but personify a place, in this case Innocent House, which houses the publishing firm that is the focus of the mystery. It is not just a place where murders have taken place; it is the cause and the answer. It has importance over all else: "For the past hundred years Pevrell Press have squandered resources on maintaining Innocent House as if the house was the firm. Lose the house and you lose the Press. Bricks and mortar elevated to a symbol, even on the writing paper" (*Original Sin,* p. 296).

Not only has the house been elevated to such a status for a hundred years, it has been the cause for murder for over a hundred years. In the end, we find that a long-believed suicide was, in fact, murder. In 1850, just before his own death, Francis Pevrell wrote the following confession:

> I killed her because I needed her money to finish the work on Innocent House . . . She loved me but she would not pass [her funds] over. She saw my love of the house as an obsession and a sin. She thought I cared more for Innocent House than for her or for our children, and she was right. (*Original Sin,* p. 495)

Dalgliesh does not always stay put in London. *Devices and Desires* is one example of James providing a plot to get Dalgliesh out of London and into the countryside. In this case, the mystery is set on the northeastern coast of Norfolk. The sojourn allows James to provide descriptions of rural England, which contrast nicely with urban England, and to provide a subplot focused on the development of nuclear power. Throughout the novel, the debate over nuclear power is a key underlying theme, the power plant itself symbolizing the dichotomy. In referring to the huge rectangular bulk of the nuclear power station, this dichotomy is captured: "But sometimes, on the darkest nights . . . both the science and the symbol would seem to him as transitory . . . and he would find himself wondering if the great hulk would one day yield to the sea" (*Devices and Desires,* p. 63).

The nuclear power plant represents the intrusion of modernity into the pristine English countryside; as well, it calls into question the whole argument about the advisability of nuclear power and the ability of technology to solve all of society's problems. The issues are so important to the novel that some have suggested that the murder is secondary to the exposition of these themes. One thing is for certain, P. D. James has had a major influence on the evolution of the police procedural into a particularly British variant of its American cousin.

Although there typically isn't the gathering of all suspects at the end of the police procedural so that the inspector can detail the solution, P. D. James cannot resist the temptation to add a bit of this Great Detective formula to the procedural variant. It begins with her first Dalgliesh novel, *Cover Her Face,* in which the body is discovered behind a locked door (although it is clear from the outset that the murderer used a ladder to escape through the bedroom window), and in which Dalgliesh gathers the suspects together in the business room while the actual events surrounding the murder are painstakingly revealed and the murderer ultimately identified. James' penchant for locked-door scenarios lasts for only one novel, but the use of extended explanations and the gathering of suspects persists for over thirty years: "Dalgliesh had said that he had wanted to see all the partners in the boardroom at three o'clock and that Miss Blackett should be with them. None of them made any objection either to the summons or to her proposed presence" (*Original Sin*, p. 472).

James has been extremely successful in transferring this American genre to Britain; yet she is unable to break away entirely from her British mystery roots. ▪▪▪▪▪▪▪▪▪▪▪▪▪▪▪▪▪▪▪▪▪▪▪▪▪▪▪▪▪▪▪

Colin Dexter's Oxford

[D]idn't it depend on exactly whereabouts on the globe one happened to be standing at any particular time?
(The Secret of Annexe 3, P. 28)

SIXTY MILES UP THE Thames from London lies the hallowed academic sanctuary of Oxford, where Colin Dexter has located his series featuring Chief Inspector Morse (we never learn what his first initial *E* stands for) and Detective Sergeant Lewis. The series has been used as the basis for a popular British television series that has been exported to the United States through the PBS *Mystery!* series, with reruns showing on the A&E cable channel. One limitation of reading this series, for television viewers, is that clear pictures of the characters Morse and Lewis have been provided so that it is difficult not to imagine John Thaw or Kevin Whately as one reads the exploits of the two Oxford detectives.

One criticism of the series has been that it is more reflective of the Great Detective tradition than the true procedural. As with James' Dalgliesh, such an appraisal does not do justice to the more complex nature of the relationship between Morse and Lewis. There are remnants of the Great Detective approach in this relationship, but more in how each one views the interaction. After Morse has ignored Lewis' advice about a lead that seems especially promising to him, Lewis ponders the nature of their relationship: "Morse was always saying that they were a team, the two of them. But they didn't function as a team at all" (*Last Seen Wearing*, p. 94).

Lewis often sees himself as a "dupe" of the chief inspector, and as often Morse deals with Lewis as if the detective sergeant were simply a "go-fer." In actual fact, they do work as a team, employing procedures tediously and meticulously, each one contributing to the resolution of the mystery. In many cases, it is Lewis who uncovers clues that allow the investigation to proceed.

Like James' Dalgliesh, Morse also represents a bridge between the traditional British Great Detective and the procedural-based police officer. On the one hand, Morse carries on the more classical tradition of British detectives of a higher ilk. An Oxford man himself, he enjoys classical music and single-malt scotch and is an accomplished crossword puzzle aficionado. Although romantic interests occur throughout the novels, he remains a not-so-confirmed bachelor. Manifold twists of plot prevent him from forming a permanent relationship with his various romantic interests. In *Last Seen Wearing,* what seems like a potentially serious relationship is terminated by the ultimate twist of plot—the lady of Morse's affections turns out to be the murderer. Unfortunately for Morse, there is both solution and resolution in this novel.

> Morse went along once more to the cell block, and spent a few minutes with Sue. "Anything you want?"
>
> There were tears in her eyes as she shook her head, and he stood by her in the cell, awkward and lost. "Inspector?"
>
> "Yes."
>
> "Perhaps you can't believe me, and it doesn't matter anyway. But . . . I loved you."
>
> Morse said nothing . . . For a while he could not trust himself to speak, and when he did he looked down at his darling girl and said only, "Goodbye, Sue." (*Last Bus to Woodstock,* p. 281)

On the other hand, Morse possesses characteristics more closely associated with police of the procedural bent—he drinks and eats too much, smokes constantly, and generally does not take himself too seriously. He also makes numerous mistakes, drawing wrong conclusions during the course of investigations. These more human traits are highlighted in some novels more than others. The setup for *The Wench Is Dead* is a hospital stay for Morse, resulting from a stomach hemorrhage. It is during this stay that we get a clear picture of the status of Morse's health as hospital staff make it clear that his eating, drinking, and smoking habits are major contributors to his health problems.

It is also in *The Wench Is Dead* where Dexter creates a different kind of plot. From his hospital bed, Morse becomes intrigued with and begins to investigate, unofficially, a hundred-year-old murder. This presents a twist in the historic murder mystery—a contemporary police officer solving a hundred-year-old crime. In and of itself, this is an entertaining plot device; additionally, it provides Dexter the venue to create a sense of life along the Oxford Canal in the mid-nineteenth century. It is through Morse's learning about mid-nineteenth-century canal travel that he becomes suspicious about the murder of a young woman and the subsequent conviction of the boatmen found responsible for the murder.

The portrait of the canal between Coventry and Oxford is supplied to Morse through the writings of another character who had examined the case in some detail. After his discharge from the hospital, Morse continues his inquiries in person while still on sick leave from the department. He visits the places involved in the murder to try to understand what it was like and what might have happened.

As in several of the Morse novels, there is a lack of total and authoritative resolution, again another variant of most police procedurals, not to mention the more traditional British mysteries. In *The Wench Is Dead,* Morse discovers that the woman thought to be murdered had not been; rather, she had staged her death for insurance money. Two men had been hanged, one imprisoned falsely. But there was nothing to be done over a century later.

In *Last Seen Wearing,* a change of identity precipitates murder during the investigation. As it turns out, the person believed dead at the beginning of the novel is not dead but becomes the murderer. In the end, Morse discovers what has happened; but by then, the murderer has escaped, and the case remains solved but unresolved. *Last Seen Wearing* is also an excellent example of Dexter's making Morse fallible. He makes numerous mistakes, following numerous leads to false conclusion, and providing several different logical resolutions of the crime, only to be proven wrong.

For sense of place, Oxford plays an important role in most of the plot lines. There are two aspects to the sense of Oxford. One is its own character as a center of education and learning. The second is its relationship to London.

The primary images of Oxford revolve around its personification as one of the world's leading academic centers. Colleges within the university, professors and staff, and students and associates are important parts of many of the plots. The fact that Morse is an Oxford man himself leads to some interesting interchanges. It also gives him a more elevated status in dealing with the various components of the university than would normally be afforded a police officer. *Last Bus to Woodstock* focuses on the university, and the intrigue that emerges in the workings of one of the colleges is an essential plot element.

To many outside of Britain, however, the images of Oxford as a major urban center where a great university resides and whose colleges are spread throughout the city is a shock. Many Americans assume that English centers of academe are nestled in a rural tranquility more closely associated with Cambridge. Oxford is an urban university. One can imagine numerous readers mimicking the surprise of one American tourist as she arrives in Oxford for the first time: "'Oxford? This is *Oxford*?'" (*The Jewel That Was Ours*, p. 11).

Last Bus to Woodstock also features the importance of the relationship between Oxford and London. The timing of trains between the two becomes a key clue to the resolution of the crime. School schedules and travel between Oxford and London are also important plot elements in *Last Seen Wearing*, although in this novel, it is not clear whether anyone has actually been killed, a fact that disturbs Morse: "'It's just not my sort of case, Lewis. I know it's not a very nice thing to say, but I just get on better when we've got a body—a body that died from unnatural causes. That's all I ask. And we haven't got a body'" (*Last Seen Wearing*, p. 114).

It is also disturbing to the genre. Although a murder will be committed later in the investigation, not knowing whether we're trying to discover clues about a murder is unsettling to the reader as well. It is another twist to the genre, responding to the challenge: what happens if a murder hasn't actually been committed at the beginning of the novel? There is a certain comfort when a murder is finally and unmistakably committed so that, like Morse, we can get on with the business of solving an actual crime.

It would be a crime not to applaud Dexter's use of maps in many of the novels. They range in complexity and quality from a sketch map of offices in *The Silent World of Nicholas Quinn* to a detailed and professionally produced map of the Oxford Canal in *The Wench Is Dead*. In all cases, they facilitate our ability to follow the inquiries of Morse and Lewis and further broaden our sense of Oxford and its surrounding environs. ■

Rural England: The Yorkshire of Peter Robinson

Gristhorpe thought for a moment.
"There might be a better way," he said at last.
"If my geography's correct."
(The Hanging Valley, P. 23)

APPARENTLY, MANY PEOPLE perceive Yorkshire as a hotbed of murder, at least fictional murder, which would explain the popularity of several police procedural series based in this part of rural England, including series by June Thomson, Reginald Hill, and Peter Robinson. All of these authors know their Yorkshire well; their senses of Yorkshire revolve around issues of historical continuity, urban intrusion, and rural resistance to changes in class-based relationships. In most cases, however, we are left without vivid images of what it looks like. Such images are more completely provided by Peter Robinson.

The Inspector Banks series has established Robinson as an exceptional craftsman when it comes to creating a sense of Yorkshire. He begins with the basic components of the police procedural—Detective Chief Inspector Alan Banks, who is happily married with children and who works as part of a team out of the Eastvale CID. The plodding nature of actual police work is emphasized throughout the series, with suggestions of cases unsolved, although seldom in the novels themselves. From this foundation, Robinson uses all of the literary devices necessary to evoke vivid images of Yorkshire, England.

Robinson is a master of narrative description. Not only does he provide numerous portraits of the dales and fells of the Yorkshire region, in one instance, he uses narrative description to provide a physical geography lesson on the formation of "hanging valleys," thus answering the question in the minds of most readers generated by the title of the book. As Gristhorpe explains to Banks: "'It's a tributary valley . . . at a right angle [where the glacier] was too small to deepen it as much as the larger one that carved out the dale itself, so it's left hanging above the main valley floor like a cross section'" (*The Hanging Valley*, p. 23).

Whereas the use of dialog furthers our impressions of places like Yorkshire, one aspect that is particularly difficult to capture is the use of dialect, quite a recognizable part of Yorkshire. Fortunately, Robinson chooses only a select number of characters to expose the patterns of dialect. A whole book filled with such characters would discourage any reader from even beginning such a burden. Interspersed sparingly, it adds to the flavor of place: "'Aye. Betty, lass, come over 'ere. T'Inspector wants a word wi' thee'" (*The Hanging Valley,* p. 223).

In effect, these characters serve as icons of local color. As with most police procedurals, however, the primary icon is the lead character, in this case, Alan Banks. Over the course of the eight-book series, we get a picture of a fairly complex individual. He is an outsider to Yorkshire, having transferred and moved to Eastvale from London several years ago, primarily to escape from the apparent disintegration of his career and personal life.

> [Banks] had transferred from London . . . He liked detective work and couldn't imagine doing anything else, but the sheer pressure of the job—unpleasant, most of it—and the growing sense of confrontation between police and citizens in the capital had got him down. For his own and his family's sake, he had made the move. (*The Hanging Valley,* p. 10)

In Banks we find society's ambivalence toward urban culture. Later, on another case, Banks gets the opportunity to return to London, where he puts to rest any question of the love-hate nature of his relationship to the big city: "So, here he was, back in London for the first time in almost three years . . . taking in the atmosphere, loving it about as much as he hated it" (*Past Reason Hated,* p. 183).

But it's not just in his ambivalent attitude toward urban society that Banks serves as an icon; he also carries many of the contradictions of society itself, as in his views on capital punishment, where he is torn between his feelings of revenge against people whose actions he feels forfeit their membership in the human race and his feelings that wanting revenge makes us no better than the criminals: "The two sides of the argument struggled for ascendancy; some days sheer outrage won out, others a kind of humanism took supremacy" (*Wednesday's Child,* pp. 228–229).

Add to all these contradictions, attention to detail, one of which is the English pub, which Robinson takes great care to describe in whichever city Banks ends up. Key here, however, is not just that vivid images of the traditional pub are provided; rather, the descriptions give a flavor of the importance of the pub in the social life of Britain, even more so in rural England. In many instances, the pub is a primary gathering place for local suspects, a place for interviews and gathering information. In

The Hanging Valley, finding the right authentic English pub in Toronto becomes the key to finding a key witness in murders that have occurred back in Swainsdale.

In keeping with the police procedural genre, great attention is paid to the whydunit, as well as the whodunit and the howdunit. As a result, numerous social issues are addressed in this series, including alcoholism, prostitution, pedophilia, psychopathy, spousal abuse, as well as the more traditional concerns with the legacy of the class system, the intrusion of urban culture into the pristine rural English countryside, and urban decay. Even in Eastvale, there is a striking contrast between the traditional idyll of British villages and the housing projects that now contain the less fortunate segments of society. Robinson calls multiple senses into play as he reveals that, even in rural England, we can find the negative aspects of urbanization.

> A desolate, stunned air pervaded the East Side Estate that morning, Banks sensed, as he walked from the mobile unit to the school . . . and those people he did see . . . had their heads bowed and seemed drawn in on themselves. He passed the maisonettes with their obscene messages scrawled on the cracked paintwork, and two blocks of flats—each fourteen storeys high—where he knew the lifts, when they worked, smelled of urine and glue. Hardly anyone was out on the street. (*Wednesday's Child,* p. 32)

This is quite a different picture of pristine rural England, one that is further underscored with recurrent references to the high unemployment rate and depressed economy of the Yorkshire region. In *The Hanging Valley,* the need to emigrate to find work, and the subsequent desire to return to Yorkshire, provide an important plot element. A reference to the policies of the Thatcher government even finds its way into the narrative. When explaining to Banks why her son stayed away in Canada for a period of time, the suspect's mother replied, "'Money. No work for him here, is there? Not with Thatcher running the country'" (*The Hanging Valley,* p. 68).

There is the implied correlation between economic depression and the encroachment of the urban into the rural, specifically targeting Leeds and York as representative of the optimistic march toward the urbanization of rural Yorkshire during the earlier, post–World War II boom. Now a very different perspective captures the sense of these urban centers.

> The Merrion centre was one of the first indoor shopping malls in Britain. Built on the northern edge of Leeds city centre in 1964, it now seems something of an antique, a monument to the heady sixties' days of slum clearance, tower blocks and council estates. (*Final Account*, p. 121)

Criticism of the deleterious effects of urbanization are not limited to the apparently failed efforts at social engineering through urban design and planning. Robinson provides a historical context for the urban blight on the serene Yorkshire countryside, beginning with nineteenth-century textile towns: "It was so easy to get lost in the conurbation of old West Yorkshire woollen towns. Built in valleys on the eastern edges of the Pennines, they seemed to overlap one another, and it was hard to tell exactly where you were" (*A Necessary End*, p. 292).

Can there be any more biting criticism of the effects of urbanization than to lose one's sense of place? I think not. If nothing else, it is through a seemingly rural-oriented series that we come to appreciate how difficult it is to avoid the ubiquitous reach of urban processes. The serenity of the English country appears to be not so serene, its rural character altered by the progress of modernization and its co-conspirator, urbanization. ■■

Glasgow and the Scotland of Peter Turnbull

> *The forty miles between Edinburgh and Glasgow represents a distance measured in more ways than a simple linear measurement: it represents a cultural gulf.*
>
> (The Killing Floor, P. 33)

FOR THE PAST FIFTEEN-PLUS YEARS, Peter Turnbull has depicted a sense of Scotland through the exploits of Glasgow's P Division, a fictitious police unit, akin to McBain's 87th Precinct. By Turnbull's own admission, he sees the series as a social commentary on Glasgow and Scotland as much as a police procedural series. Like

McBain's series, the strength of the series begins with detailed character descriptions, in this case including numerous cops, instead of a single main character throughout, and detailed police procedures.

In *The Killing Floor,* no fewer than half a dozen cops, spread out over all three shifts, conduct the investigation. The body is found on the graveyard shift, which is being worked by Detective Sergeant Sussock. Detective Constable Montgomerie holds down the swing shift, while DCs King, Abernethy, and Wilhelms work the day shift. The overall investigation is coordinated by Detective Inspector Donoghue. As a result, continuous investigation provides a great deal of activity and the resolution of the crime within a week of the discovery of the body. Adjunct personnel, such as the pathologist and the forensic chemist, whose characters are also revealed in some detail, add to the feeling of a true team effort in the P Division bringing justice to Glasgow.

For purposes of sense of place, several additional characteristics emerge as strengths of the series. First is Turnbull's attention to contrast, contrasting the Scottish countryside with urban Glasgow and contrasting the different sections of the city itself.

We are reminded that much of Scotland is countryside and that between the major urban centers, one finds the glorious countryside we often associate with the northern British Isles:

> and beyond that was a field of lush pasture, and beyond that a second similar field but of higher elevation, and beyond that was rough grazing, then the hills, and beyond the hills the wilderness which stretched forty miles to the suburbs of Edinburgh. (*Long Day Monday,* p. 8)

Yet, even within the city itself, one can find rather idyllic parts: "a small quiet enclave, a few tree-lined streets which had a 'village feel' about them, despite being in easy walking distance of the city centre" (*Long Day Monday,* p. 54).

These scenes, however, clash starkly with those parts of Glasgow where one is reminded that much of the British north is economically depressed and that urban squalor and deprivation have found Scotland as well, especially acute in Glasgow, which is the industrial center of Scotland. In *The Killing Floor,* the motive for murder is related to a substandard housing project located in one of the low-rent districts of Glasgow. Contrasts are made continually between a high-rent neighborhood, where the body of a social worker is found, and the low-rent neighborhoods, where she worked and where her clients lived.

A sharp contrast is drawn between industrial Glasgow and the Scottish capital of Edinburgh, where the author is able to explain how mileage is only one way to measure distance, and that physical separation is only one way to assess differences between these two dominant Scottish cities: "Despite the short traveling time, Donoghue, and many like him, felt massively distanced from one when in the other" (*The Killing Floor,* p. 33).

And then there is the tie to England. Given sarcastically is a description of the way to London from Glasgow. The metaphor reminds us of the long-bitter resentments many Scots hold about their dependent relationship on England as part of Great Britain: "some two miles distant, was the thundering A74; the 'big road', the 'trap-door' route to England. By far the quickest way to leave Glasgow for England by car: open the trap door, and fall south. It takes two hours" (*Long Day Monday,* p. 9).

Another distinguishing feature of the Scottish countryside is the use of colloquial speech, which Turnbull, much like Robinson in Yorkshire, uses sparingly but effectively. One additional device, however, is the lack of speech. In contrasting urban and rural Scottish lifestyles, the novels are sensitive to the true nature of rural folk. Characters do not have to talk all the time to provide authentic dialog.

> [The policeman] realized that [the farmer] wasn't staring at him; he was communicating with him, and he was communicating a message which read, "Something is amiss . . ." without speaking and, falling in with the ways of non-verbal communication of the people of the land of the red roads, he walked up the verge. (*Long Day Monday,* p. 15)

The physical geography also comes into play, not just as setting for the crimes, which it is, but also because it facilitates the resolution of the mystery. As a result of Scotland's northerly location, not only is Turnbull able to provide a lesson on the effects of relative location, but he also provides his police with increased opportunities for detection: "'We're into the white nights, I've got daylight until 22:30, I'd like to sweep local open areas'" (*Long Day Monday,* p. 27).

But it's not just physical geography that plays a role. Knowledge of spatial relationships is key as well. Using the ubiquitous wall map in the department—ubiquitous but not always referred to in mystery novels—the police are able to eliminate certain parts of the city to look for a missing child and to focus on others. A grasp of spatial relationships also helps to eliminate some suspects, while focusing on others: "'because the location of the graves is too far apart to suggest local knowledge. They are the same general area, that's all'" (*Long Day Monday*, p. 198).

Where would the detectives be without geography? Like all of us, they'd be nowhere, and their cases would go nowhere. Peter Turnbull's writing is an excellent example of blending the importance of place with a scintillating mystery. As with numerous other authors in this study, Turnbull also places an emphasis on the need for the investigator to "get a feel" for the world of the murdered and the murderer, to become immersed in the places associated with the crime. Once again the physical environment and the psychological environment intertwine. Several characters, including Detective Inspector Donoghue, comment on the strong sense of coldness and oppression generated by the house where the first killing occurred.

> [I]t was the coldness of the huge house in Pollockshields, on Sherbrooke Avenue, which stayed with him and remained with him for many years. It was not simply a matter of temperature, it was a coldness mixed with a sense of oppression, a "presence," a sensation of the building itself being hostile. (*The Killing Floor*, pp. 35–36)

At one point, Turnbull cannot resist blending the real with the fictional. Just as one of the basic characteristics of all police procedurals is the suggestion that many cases go unsolved, we often find references to other police procedurals or at least to the heroes of the procedurals, identifying them clearly as fictional and thus lending a sense of reality to the current mystery. Turnbull chooses a classic:

> "There is a technique for rebuilding the face and head using the skull as a base, it was pioneered by a Soviet scientist. You may have seen the film *Gorky Park*."
>
> "I have, yes."
>
> "The technique featured in that film, it has been adopted and is used in the UK." (*Long Day Monday*, p. 119)

The blending of references to other works of fiction and the suggestion of reality adds credibility to the story line and serves as a reminder that the boundary between the two is often indistinct. ■ ■ ■ ■ ■ ■

Dublin and the Ireland of Bartholomew Gill

It is what invites but will not submit to description in simple words—the cold, wild beauty of Ireland.
(The Death of Love, P. 72)

THE JOURNEY around the British Isles concludes outside of Great Britain, on the Emerald Isle of Ireland. Once again we return to a series that features a single lead character—Chief Superintendent Peter McGarr of the Serious Crime Unit (Murder Squad) of the Garda Siochana (Irish Police) out of Dublin. Gill brings a particularly Irish slant to the genre in all of the devices he employs. We are treated to narrative descriptions of the Irish countryside, as well as the urban areas, from the perspective of one who has a deep love for the land, as well as a healthy respect for its positive and negative features.

A sense of Ireland begins with a description of the landscape. The description takes on a decidedly contrasting hue for Gill—the contrast between the modernized, urbanized Dublin and the pristine, Irish countryside. As we would hope and suspect, the Irish countryside, which has inspired writers and poets for centuries, cannot simply be described. Even in escapist literature, it must be commented upon, often personified, for its effect on the human psyche, a psyche that demands constant references to being Irish: "Good traditional Irish music touched something deep in Bresnahan. Having been preserved with fierce racial pride, it was a mixture—curious in this day and age—of life and antiquity" (*The Death of Love*, p. 203).

We find in the McGarr series, however, that being Irish is a far more complicated issue than the homogenized characterization we often get outside of Ireland. *The Death of an Irish Tinker* focuses on one aspect of this complexity, where we find victims and potential victims are Tinkers (also referred to as "Travelers"), simply defined as the Irish equivalent of gypsies. Coming from a fine itinerant tradition, serving as seasonal labor, horse traders, minstrels, storytellers, thatchers, and chimney sweeps, that dates back to the fifth century, their pattern of life changed in post–World War II Ireland

> with the appearance of cheap metal and plastic goods and mechanized farming that replaced seasonal labor. Forced to shift to

> cities, Travelers were cramped onto small plots heaped with rusting auto parts, trash, and such. The men went on the dole; the women begged in the streets. (*The Death of an Irish Tinker*, p. 43)

Yet, Gill's characterization of these people is sympathetic, humanizing even. He comments on their sense of community and contrasts it with the coldness of more legitimate Irish society. Forced to flee to the safety of a Tinker camp, a young female Tinker, who has been raised in high society and never exposed to the Tinker way of life, is struck by the contrast between these societies.

> Oney Maugham immediately took to life in a Traveler camp. After the distant, impersonal, up-market neighborhoods of Reigate and Ballsbridge, where she had passed most of her girlhood years, the warmth of the greeting that her grandparents and she received at the Traveler encampment in Cork seemed magical. (*The Death of an Irish Tinker*, p. 178)

Another part of the complex issue of being Irish is captured in the character of McGarr himself, who is often used to address the issue of being Irish, and his family, which is used to epitomize the contrast of social status in Ireland, this time between Protestants and Catholics: "[McGarr's wife Noreen] the only child of landed Protestant gentry who concerned themselves with horses, fine art, and the conservation of principal; he the fifth son of a Catholic Guinness Brewery worker" (*The Death of Love*, p. 10).

Many of the characters in the series belong to large families, including McGarr himself who is one of eight children. The McGarr character is also interesting from the perspective of his appearance and the contrast between male and female authors writing about male protagonists. In contrast to female authors who generally create strikingly handsome lead characters, male authors such as Gill create lead characters of no particular physical appeal.

> [McGarr was] a short, well-built man of fifty-five with an aquiline nose that had been broken more than once and was now set to one side. The hair that could be seen beneath a waterproof fedora was red and curled at the back of his neck. Otherwise he was bald. (*Death of an Ardent Bibliophile*, p. 7)

Unfortunately, another aspect of being Irish relates to the slumping status of the Irish economy, which is brought to our attention throughout the novel, especially as Ireland prepares for its role in the European economy: "With unemployment over 20 percent, young people—especially graduates from Ireland's several, fine universities—had to emigrate to find work. Over thirty thousand had left in the last year" (*The Death of Love,* p. 18).

Pivotal to the murder in *The Death of Love* is the pending sale of controlling interest in the Irish bank, The Bank of Eire, to Japanese interests. It is particularly relevant because of the Japanese desire not to get shut out of the post-1992 European economy: "Eire Bank will give you [the Japanese] what you need, an unfettered toehold in post-'92 Europe. It will also get you continued government support" (*The Death of Love,* p. 212).

Dublin and Dubliners are highlighted throughout the series, as we are provided with a definition of the meaning of the word *Dublin,* which is both historical and metaphorical. It also gives an opportunity for a brief physical geography lesson.

> *Dubh Linn.* In Irish it meant Black Pool. The surrounding mountains and hills formed a bowl around the city, and once in a great while . . . a sudden, chill west wind off the Irish Sea established an atmospheric inversion. Combined with urban air pollution and the smoke from the hearth fires that Irish people still preferred, its effect was stunning and foul. (*Death of an Ardent Bibliophile,* pp. 3–4)

Just as the urban character of the capital city is captured in narrative description, the particular dialect of the capital is captured through the use of colloquial dialog.

> [S]he made her voice low, gravelly, and pure Dublin. "Oi wanted teh phone yiz, sure I did. But it was too late. Und 'dere for the longest time. I kept t'inking Oi was coming home directly." (*Death of an Ardent Bibliophile,* p. 26)

> "Yeh're tryin' to floy," she said in the pancake accent that marked her as a Dubliner born and bred. (*Death of an Ardent Bibliophile,* p. 30)

The ever-present specter of the IRA does find its way into the plot. And although it is important in facilitating the final act of murder, it does not dominate or detract from the primary focus of the mystery, which is not directly related to the IRA. Yet, it is a reminder of the constant antagonism that exists in Irish society, not solely in the north.

Even though the McGarr series is replete with geography and sense of place, it doesn't hurt to have an unsolicited testimonial for the vale of geography.

> It was twenty-five or so steep, sinuous miles to Waterville alone, where Gladden would then have to choose between shore or mountain routes, which were some thirty-five or forty additional miles long, if McGarr's knowledge of geography served him well. (*The Death of Love*, p. 267)

. . . which, of course, it did. ■■■■■■■■■■■■■■■■■■■■■■■■

Conclusion

These examples of the British response to the police procedural format are the result of answering the challenge: what happens if you blend the procedural approach with aspects of the traditional British whodunit? The answer is increased variability. Just as some British authors have made the transition from the classical detective genre to the police procedural, they have also expanded their use of place—from primarily landscape descriptions to more detailed conveyance of all the physical and cultural characteristics of the locales. The English response tends toward interesting protagonists with certain traits carried over from the Great Detective tradition. The English also tend to emphasize the antagonisms between urban areas and the countryside. The Scottish and Irish variants produce police officers more in keeping with the American example, and they also emphasize strong senses of nationalism, which often focuses on hatred toward the English. As we travel to the European continent, we will find series that borrow from both the American and the British variants of the police procedural.

Murder on the European Continent

Wahloo & Sjowall

Van de Wetering

Freeling

Dibdin

Chapter 5

For many mystery fans, first memories of murder on the European continent trace back to the Orient Express and the famous Hercule Poirot, notwithstanding the fact that the best-known Belgian detective was the creation of a traditional British mystery writer, Dame Agatha Christie. Not surprisingly for the English-language world, many early attempts at crime writing about places continental were penned by British and American authors, a tendency that continues to this day. Several authors have adapted the police procedural genre quite nicely to the continent, and four of these have been selected for this book as representative of this region of the world. Two of the four are not native to the regions of their crimes, while the other two are. The former are Nicholas Freeling, who writes about provincial northern France, and Michael Dibdin, whose crimes take place in Italy. Freeling was born in London, raised in England and France, and lived many years in France. Although Dibdin is British, he spent four years teaching at an Italian university. Thus, both non-native writers have solid credentials for the cultures that provide the settings for their novels. Our native authors are Janwillem van de Wetering, who writes about Amsterdam, and the husband-wife team of Per Wahloo and Maj Sjowall, whose crimes occur in Stockholm.

Conspicuous by their absence are three enormously popular series written by Georges Simenon, Mark Hebden, and Magdalene Nabb. Simenon and Hebden fit squarely into the tradition of the Great Detective genre. The Nabb series contains elements of the police procedural. However, unlike Dibdin's work, which is included, places in Nabb's books serve only as settings; they do not become essential plot elements. Thus, the series does not meet the criteria for being defined as a place-based police procedural.

The four series selected are place-based and offer a diverse selection of geographical regions. We begin in the south and move north, starting with Michael Dibdin's Aurelio Zen, based in Rome. From Italy we venture to northern France to encounter Nicolas Freeling's Henri Castang. Freeling authors another series based in Amsterdam, which was not chosen because we have another series based there, featuring Janwillem van de Wetering's Detectives Grijpstra and de Gier. Even farther to the north, Per Wahloo and Maj Sjowall base their Martin Beck series out of Stockholm. In all of these series, the sense of place is not only conveyed strongly, it is essential to the plots of the novels. ■ ■ ■ ■ ■ ■ ■

The Italy of Michael Dibdin

"What you want, signore, this famous 'order' of yours, is something un-Italian, un-Mediterranean. It's an idea of the North, and that's where it should stay."
(Ratking, P. 11)

MICHAEL DIBDIN PROVIDES IMAGES of Italy through the exploits of Vice Questore Aurelio Zen of the elite Criminapol squad of the Italian Ministry of the Interior. When we first encounter Aurelio in *Ratking,* he is Commissioner Zen, having been demoted to desk duties some years earlier because of his handling of the Moro kidnapping. Later, it is revealed that his banishment resulted from being too efficient and too close to uncovering embarrassing complicities of the government and police. Another kidnapping, this time in Perugia, in northern Italy, and the demand for a high-profile inspector from Rome, provides Aurelio the opportunity to regain favor, which he does along with promotion to vice questore.

The recurring themes of the Zen series are introduced in *Ratking*—the duplicity of the Italian legal system, where the line between right and wrong often is blurred, and the need for a member of the national police to travel throughout the country to solve crimes. From Zen's home base in Rome, the series visits the aforementioned Perugia, his native Venice, Milan, and Naples. We can add the Vatican, which is just "like any other foreign country," to the list of other locales for murder.

It is in Aurelio's travels that we become aware of the regional differences and animosities that exist in Italy. Clearly the northern feeling of superiority over southerners appears often. As well, the use of his current home in Rome, which he refers to as "the city of his exile," compared to his native Venice, highlights the way in which the capital city is perceived. Aurelio is never comfortable in Rome.

> [Aurelio] would never learn to like Rome, never be at ease with the weight of centuries of power and corruption there in the dead center of Italy, the symbol and source of its stagnation. How could he ever feel at home in the heaviest of all cities when he had been born and formed in its living antithesis, a city so light it seemed to float? (*Ratking,* p. 12)

Narrative description is clearly one of Dibdin's strengths, where he incorporates attention to detail to such an extent that one gets a feel for Rome, Venice, or Milan. In Venice, for example in *Dead Lagoon,* the dominating presence of canals is essential. To encourage visualizing the nature of the canal-oriented layout of Venice, Dibdin instantly alerts the reader to its maze-like nature by providing two, high-quality maps of different scales with distance legends. These prove useful as one tries to plot, excuse the pun, Zen's movements throughout *Dead Lagoon* and to envision the spatial relationship of places in and around Venice.

Yet the layout of the city as captured on any map cannot allude sufficiently to the true nature of narrow streets opening onto piazzas of various sizes and shapes. For this, Dibdin incorporates a keen attention to detail and numerous glimpses of the city as Zen goes about his investigations.

> Zen turned left into an alley so narrow he had to walk sideways, like a crab. It seemed to come to a dead end at a small canal, but at the last moment a portico was revealed, leading to a bridge . . . An assortment of débris was visible at the bottom of the water: the wheel of a pram, a punctured bucket, a boot. A large rat slithered across the mud and hopped into an open drain. In older buildings, people still kept a heavy stone on the toilet cover to stop the creatures from getting loose in the house. (*Dead Lagoon,* p. 28)

Physical geography finds its way into the plot in all sorts of ingenious ways. In *Cabal,* Dibdin uses a trip to Milan to provide a rather detailed description of the Apennine mountains, with numerous brief references throughout the trip, and the history of man's attempt to pierce the range: "The Apennines form a continuous barrier running almost the entire length of the Italian peninsula" (*Cabal,* p. 215).

As essential as topography, climatology, religion, and politics are to understanding contemporary Italy, so is history, of which Italy has a plentiful amount. Dibdin approaches history in two ways—from a background point of view, in which history is used to help explain the why, and from the perspective of personal history, most often through the eyes of Aurelio Zen himself. In *Cabal,* the history of the Church, and particularly of certain important brotherhoods like the Knights of Malta, comes into play. In Venice, it's the history of Nazi occupation, treatment of the Jews, and the impact it had on Aurelio's family and friends that helps to uncover the truth.

At the same time, changes during his own lifetime bring to the surface key issues, one of which is the impact of modernization and urbanization, for this is first and foremost an urban series. Although many things have changed in his native Venice as a result of economic changes and urbanization, some aspects of the urban environment remain the same, including

> the east end of the city, the maze of former slums crushed in between the Pietà canal and the high fortress walls of the Arsenale. This was a secretive and impenetrable district, of no particular interest in itself and on the way to nowhere else. (*Dead Lagoon,* p. 112)

And to ensure that the importance of historical continuity is not lost in the complex development of plot, Zen's reaction to the successive attempts at grandiose construction projects in Rome, even from the time of Nero, leaves little doubt: "This stirring historical perspective, far from inspiring Aurelio Zen to a sense of wonder and pride, merely intensified his oppressive conviction that nothing ever changed" (*Cabal,* p. 175).

His convictions help form the sense of place portrayed in this series. We buy into his convictions because in Aurelio Zen, Michael Dibdin has created a character we can relate to on a personal level. Not only does Dibdin take great pains to develop this character, he also intertwines Zen's personal life with his professional cases.

One of the changes that merits special attention from Aurelio Zen is the expansion of the Rome metro system, which gives Dibdin the opportunity to present the ever-present antagonisms between continuity and change. The metro system represents innovation "in order to deal with the worsening crisis of traffic movement," or lack thereof, in Rome, a situation Dibdin refers to at numerous opportunities. Yet Zen is resistant to these changes to the point that instead of using the new metro line, which was convenient and much faster, he continues to commute to work by bus: "The bus journey was by no means an unrelieved joy, but it took place in a real city rather than that phantasmagoric subterranean realm of dismal leaky caverns that might equally well be in London, or New York—or indeed the next century" (*Cabal,* p. 162).

Not only does Zen represent resistance to change, he also makes a nice case for the aesthetic value of place. He is not alone in this conviction, nor in his resistance to the encroachment of the metro system, as we shall see as we venture to Amsterdam later in this chapter.

Unlike most police procedurals, in this series Aurelio Zen is not part of a single team. As vice questore (the Italian equivalent of detective chief inspector), his day-to-day involvement, and frustration, with the Italian police bureaucracy is omnipresent. But his role as part of a team is not a primary focus of his procedures. This doesn't mean that he works alone; rather, different teams are put together for each of his novels, especially when he ventures outside of Rome. And although his own criticism of police inefficiency extends to all places he's assigned, he sees the public's lack of respect for the police as a strength: "'We have an unfair advantage in the police. Everyone assumes we're stupid'" (*Ratking,* p. 222).

Weaknesses in his personality make him vulnerable as a character. In *Cabal,* he is on the verge of taking a bribe at one point in order to have the money to compete with a suspected rival for the affections of his mistress. An untimely suicide precludes us from ever knowing for sure if he would have gone through with it. But the uncertainty makes Zen just a little more human, especially in a society where graft and corruption are so rampant.

Many of our glimpses into Zen's personality emanate from his relationships with two women—his mother and his mistress. It is not just that these relationships reveal the humanity of Zen, they are also interwoven into the primary mysteries. It is not all that difficult to relate to the trials and tribulations of a divorced, middle-aged man, smitten with a beautiful, younger widow. Zen becomes an imperfect, yet likable character who we can trust to interpret much of what we experience in Italy. ■■■■■■■■■■■■■■■■■■■■■■■■■■■■■■■■■■■■■

The Provincial France of Nicolas Freeling

Like all Frenchmen he was a complicator;
[his Czech wife Vera] was a simplifier.
(Wolfnight, p. 119)

ALTHOUGH FREELING MAY be better known for his Amsterdam-based series featuring Inspector Van der Valk, his series featuring Inspector Henri Castang offers an insightful glimpse into provincial northern France. One twist to the series is his noticeable anti-Paris bias and the use of the negative aspects of Parisian society as a point of comparison for better appreciating the benefits of life in provincial France. Castang also brings a rather critical, later turning to outright cynical, interpretation of the French system of justice.

The actual location of the fictional home of the Serious Crimes Brigade of the Police Judiciare (PJ) is never pinpointed, only that it is "a provincial city in France, of something under half a million souls" (*Wolfnight,* p. 1). Our protagonist, Inspector Henri Castang, who works his way through the ranks of the PJ during the twenty-year run of the series, is clearly provincial at heart and satisfied with his post. He is happily married to Vera, a defector from the Czech gymnastics team. One of the strengths of the series is the interplay between the two, with Vera bringing a very different perspective to the cases Castang investigates. On occasion, she and their children, born during the middle novels of the series, get dragged into the case, as we find in *Wolfnight.* A relatively stable relationship focuses on her role in their lives: "Vera. This young woman. Pleasant young woman, useful adjunct, complement, what-have-you, in the cop's wearisome and generally boring existence. An alibi. A comfort mechanism" (*Wolfnight,* p. 119).

As with all series in this genre, the main characters are an important part of its success. In this case, the main characters are Henri and Vera, although we also get to know Henri's boss, Commissaire Richard, as well. Whereas Vera offers a complement and counterbalance to Henri's personality, the commissaire offers a stereotype of the professional French provincial police officer, juxtaposing Castang's somewhat unorthodox approach to police work. For Freeling, character development holds a central place in the evolution of the plot. It has been suggested, in fact, that the primary focus of a Freeling mystery is not whodunit, which often we discover long before the final pages of the novel, but whydunit. Each participant, each possible suspect, each secondary character is examined in detail as the resolution of the murder appears to be used as an excuse to examine human character and to ask fundamental, philosophical questions about the human condition.

With respect to the human condition, closely related themes that recur throughout are French nationalism and racism, especially French racism toward other European cultures. Often Vera's Slovakian roots provide the springboard into these examinations. Yet the French sense superiority is not limited to Czechs and Slovaks; it includes European cultures of diverse persuasions—Poles, Jews, Irish, and even Belgians: "But [Belgium] only exists in a variety of French wit, which holds that Belgians are unbearably dense and primitive" (*Flanders Sky,* p. 7).

It is not only French nationalism and racism that gets critiqued, however. There is a strong antinationalist tone throughout the series: "The place where we were born, which made and shaped us, and whose air and soil and water we absorbed in childhood, gets muddled with the flag-waving sentimentalities of nationalist patriotism" (*Flanders Sky,* p. 77).

The British and its empire come under close scrutiny, especially as their racism takes on tones of anti-French sentiment: " 'But for an English audience a really foul villain, painted extra black because in fact he's singularly unconvincing, simply had to be French' " (*Castang's City,* pp. 210–211).

The one overridingly unique contribution of this series to the police procedural genre is the exposure it provides to the French judicial system, based on the Code Napoleon, and the role of the police within this system. It is quite different from all others discussed in this study. The initial investigation determines whether or not a "presumed" person should stand trial. It is an "inquisitorial" rather than "accusatorial" process that is directed by a judge of instruction who sets the exact nature of the investigation by the PJ. If the evidence appears to be sufficient, the dossier is presented to the Chamber of Accusation (a kind of judicial committee) to ensure that the instructions were carried out correctly. If so, the "presumed" becomes the "accused" as the case is sent to the Assize Court for trial. The main difference from the perspective of police procedures is that the limits of the investigation are set by the judge of instruction and that the police investigation is concerned more with the person than the fact, the criminal than the crime. "An instruction is strictly neutral . . . [i]nclines actually toward the defense" (*The Bugles Blowing,* p. 185). From the French perspective, although laborious and time-consuming, the French judicial system is superior to its counterparts in England or the United States because "shocking miscarriages of justice, such as have been far too frequent in the Anglo-Saxon system, are avoided" (*The Bugles Blowing,* p. 78).

One additional area in which this series excels is in the explicit appreciation for the importance of sense of place—not only in the structure of the novel, but also in the development of the plot, specifically its use by Castang to aid in resolving the mystery. In the early stages of many of the novels, Castang spends a good deal of time steeping himself in the sense of the murder place as a preliminary step in solving the crime.

In *The Bugles Blowing,* for example, after the confessed murderer of his wife, daughter, and the wife's and daughter's lover has been questioned and jailed, Castang returns to the suspect's house, where the murders took place, to spend time getting a feel for the people who lived and died here. Certain incongruities between the way the family lived and the way the three were murdered alert Castang, and the readers, that this is much more complex and complicated an investigation than is apparent at first.

The key point is the assumption that to solve the mystery, the investigator needs to understand the victims and the murderer, and that one clue to understanding people is to understand the places where they lived and died—place as an expression of their personalities reveals insight into character and possible motive. Immediately, we are reminded of James, Robinson, and Turnbull, in particular. Tied closely to this theme is the emphasis Freeling places on the disruptive, destructive nature of crime for those involved, as well as for society as a whole.

As one literary device for conveying sense of place, the use of dialog and monolog in the Castang series is at times both disconcerting and effective. Sometimes the plot is difficult to follow because Freeling often employs streams of consciousness and random thoughts, throwing in literary and historical allusions at will, rather than relying solely on cohesive conversations and logical presentations of ideas. The overall realism is compelling, if not a bit confusing, reassuring us that we are dealing with human characters with whom we can empathize, if not fully understand. To distinguish between the reality of the police procedural and fictionality of the Great Detective genre, Castang refers to Simenon's Maigret series from time to time: "it was probably the only point of resemblance between him and the best-known fictional cop in the whole world, whom every French cop has cursed heartily upon occasion" (*Castang's City*, p. 31).

Freeling brings the series to an end with *A Dwarf Kingdom* in 1996. After several years as a consultant to the European Community in Bruxelles (Brussels), Henri and Vera retire to the Atlantic seaport of Biarrits, where one last murder investigation reinforces that after twenty-plus years, Henri is ready to retire, even though we have lost an effective interpreter of northern France. Fittingly, Freeling does not feel comfortable using Henri or Vera to tell us that it is at an end. He leaves that for an epilog: "It must stay in my own mouth. But I am sorry to say goodbye. Two good friends are here, and across the last twenty-five years" (Epilog to *A Dwarf Kingdom*). If readers become attached to fictional characters over the years, it must certainly be more so for the author. ■ ■ ■ ■ ■ ■ ■

The Amsterdam of Janwillem van de Wetering

"You speak very good English you know."
"Most Dutchmen do. We have to; this is a small country in a big world and nobody speaks Dutch, except us."
(Tumbleweed, p. 40)

ALTHOUGH CRITICS CITE CHARACTER development and his aesthetic and moral insight as the true virtues of van de Wetering's series, one cannot underestimate its ability to provide a sense of Amsterdam over the course of at least a dozen novels. The series features Adjutant Detective Henk Grijpstra and Detective Sergeant Rinus de Gier of the Amsterdam Municipal Police Department. The commissaris, the senior chief of detectives, also plays a major role in the series, although the basic team is Grijpstra and de Gier.

Even the philosophical questions, however, sometimes have a geographical perspective to them as in the case of the commissaris' interview of a suspect.

> "Why are we here?" the commissaris asked.
>
> "Here in jail, you mean? Why are we here together, just? In this cell?"
>
> The commissaris smiled.
>
> "No," the commissaris said. "[This] riddle is more difficult."
>
> "You mean why are we here, you and I, on Earth?"
>
> The commissaris smiled.
>
> "It is only when I began to read science fiction that I understood something about the riddle. I didn't solve it, of course, but I knew that there is a riddle. What the hell is all this life doing on a little round ball, suspended in space?" (*The Corpse on the Dike,* pp. 217–218)

This is the kind of writing, reflective of van de Wetering's years of studying Zen in Japan, that has made the series so popular and so appealing to academic analysis. Yet van de Wetering's craft does not feature only philosophical debates. Several recurrent themes run throughout the series.

One theme focuses on descriptions of Amsterdam. The stories are generally sprinkled with images of the city—canals, dikes, islands, houseboats, people fishing, public transportation, and more are added fleetingly in many instances as our Amsterdam cops search out suspects and clues. Yet, as the series progresses, more detailed images are provided, first of the traditional Amsterdam of the postcard variety.

> The Straight Tree Ditch is a narrow canal flanked by two narrow quays and shadowed by lines of elm trees which . . . filtered the light through their haze of fresh pale green leaves. Its lovely old houses, supporting each other in their great age, mirror themselves in the canal's water, and any tourist who strays off the beaten track and suddenly finds himself in the centuries-

> old peace of this secluded spot will agree that Amsterdam has a genuine claim to beauty. (*Death of a Hawker*, p. 19)

This idyllic scene is counterbalanced by the new and the modern, not in a particularly positive light: "They were in the southern part of Amsterdam now, and gigantic stone-and-steel structures blocked the sky, like enormous bricks dotted with small holes" (*Death of a Hawker*, p. 94).

If nothing else, the ability of light to penetrate these two environments suggests a particularly resistant commentary on the advances of modern urbanization. The tone of this commentary is carried over into the subplot in *The Death of a Hawker*, in which van de Wetering describes the protests of one Newmarket neighborhood in the city against the decision to locate a metro station in the neighborhood. Reminiscent of Aurelio Zen's resistance, van de Wetering has now employed an entire neighborhood, which requires riot police to come out in force to protect the workers and the machinery, to get his point across. Yet van de Wetering's approach to many of these changes comes across as ambivalent. Redevelopment of Amsterdam is described in such a way as to suggest acceptance and acquiescence.

> But the city had come to life again and cared about itself. [Bickers] Island had been rediscovered . . . and gradually the warehouses were being restored, the gardens cleaned up and replanted with shrubs and trees, and the canals dredged. It was still a quiet place, however, for it was out of the way; one could walk about in peace and concentration. (*The Corpse on the Dike*, p. 108)

Here, van de Wetering combines positive aspects of change with the search for quiet, peaceful places, set against the onslaught of the urban environment.

Throughout all the mayhem that accompanies this urban onslaught, including murder, our heroes continually comment about how nothing ever happens in Amsterdam; how quiet and peaceful it is. These observations usually occur at the beginning of the novels, just before all hell breaks loose: "'Have you noticed,' [Grijpstra] asked hoarsely, 'that nothing ever happens in Amsterdam?'" (*Tumbleweed*, p. 1) and "Amsterdam is a quiet town. A nice quiet town" (*The Corpse on the Dike*, p. 23).

As if to fulfill this perception, we follow our Amsterdam cops as they find places of peace, often in private gardens and individual apartments, against the turmoil and mayhem of urbanization and modernization.

At the same time that the series focuses on Amsterdam, it does have something to say about being Dutch, in general a fate shared by many Europeans living in countries with small populations. The ability to speak a number of languages is a characteristic of many small European countries, where it becomes a basic survival skill.

Two specific aspects of police work are emphasized throughout the series—teamwork and the value of understanding local culture. Although the utility of these aspects is reinforced by the manner in which investigations are carried out, van de Wetering ensures that we don't miss the point: "'What I like about the police,' de Gier said, 'is our teamwork'" (*Tumbleweed,* p. 26).

Succinct and to the point right from the start, the interaction of our three principal investigators, along with numerous supporting members of the team, works to prove just that and to emphasize this basic characteristic of the police procedural. Equally as important is the understanding of local culture—in Amsterdam, this is not a problem because all three officers are Dutch. When the commissaris travels to Curaçao as part of an investigation, however, the local policeman, Silva, of Dutch descent and education, takes the opportunity to reinforce the point.

> "But you are accepted?"
>
> "I am from the island," Silva said, "born and bred. I was brought up with cabryt milk and rum. I speak the language. If I didn't, I would never solve a single crime." (*Tumbleweed,* pp. 111–112)

More exotic places also weave their ways into novels—the United States in *The Main Massacre,* and only fittingly, Japan in *The Japanese Corpse.* Much of the latter novel introduces us to Japanese culture from the perspective of a northern European, and it is a very sympathetic interpretation: "and suddenly he felt a wave of love for the hundred million people of these islands and their childlike ability to enjoy, to play the supreme game, to accept and admire the beauty of the creation and to try to live in harmony with it" (*The Japanese Corpse,* p. 107).

But it isn't just an opportunity to wax poetic about a country and people he obviously has deep affection for. Van de Wetering also uses this story to remind us of the historical links between the Dutch and the Japanese, links that go back before Meiji and that were helpful to the Dutch, but essential to the later modernization of Japan. This special relationship dates back to 1635: "we were the only Western nation allowed to trade with Japan in those days. The Japanese figured that we weren't going to convert them to anything, but were only there for the money. And so we were" (*The Japanese Corpse,* p. 55).

There is historical context for van de Wetering's feelings of admiration and awe for Japanese culture. Numerous innuendoes make it clear that the author sees Western intrusion in the Far East as something less than a mutually beneficial relationship. Could it be, perhaps, that the current drug trade to northern Europe from China through Japan is payback for centuries of exploitation? "'Yes,' he said. 'I care about the drugs. The yakusa are helping the Chinese to get even. Once the Western nations poisoned China with opium; now it is the other way around'" (*The Japanese Corpse,* p. 109).

With an emphasis on the aesthetic and metaphysical, Janwillem van de Wetering has managed to incorporate sense of place into his Amsterdam cops series, providing a slightly different approach to the police procedural, but nonetheless one that is both pleasurable and insightful into this part of northern Europe. Yet, like Freeling, it appears that van de Wetering is bringing the series to an end. *Just a Corpse at Twilight* finds de Gier retired in the United States and Grijpstra retired from the force, working as a private investigator. The police procedural component of the series, if nothing else, has been brought to closure. ▪▪▪▪▪▪▪▪▪

The Stockholm of Per Wahloo and Maj Sjowall

[A]ll varieties of crime flourished better than ever in the fertile topsoil provided by the welfare state.

(Murder at the Savoy, P. 82)

THE STOCKHOLM-BASED series of Wahloo and Sjowall represents a shift in emphasis from most other police procedurals. Whereas most authors of this genre identify pleasure and escapism as the primary goals of their stories, Wahloo and Sjowall, both avowed communists, used their police procedural series as a forum for critiquing the ideological and moral underpinnings of the bourgeois welfare state that had developed in Sweden. Ten novels, written between 1967 and 1975, complete the series, focusing on the exploits of Chief Inspector Martin Beck, who is part of the homicide detective team. The series introduces numerous permanent members of the team, including Detectives Kollberg, Larsson, Mansson, and Ronn, often spending numerous pages following members of the team other than Beck. Permanent cast members even include the radio patrolmen, Kristiansson and Kvast, further emphasizing the teamwork aspect of the police procedural. As one would expect with such an emphasis on teamwork, there is also an emphasis on detailed police procedures: "over three hundred tips had already come in. Each item of information was registered and examined by a special working group, after which the results were studied in detail" (*The Man on the Balcony,* p. 47).

An emphasis on critiquing the bourgeois welfare system does not preclude a description, albeit an ideologically charged one, of the changes that have affected Stockholm from some indeterminate idyllic past as a result of the weaknesses of the current political and economic system. This critique hits close to home as Beck painfully watches his mother live out her final years in a state-run retirement home.

> Nowadays they were called . . . "pensioners' hotels," to gloss over the fact that in practice most people weren't there voluntarily, but had quite simply been condemned to it by a so-called Welfare State that no longer wished to know about them. It was a cruel sentence, and the crime was being too old. As a worn-out cog in the social machine, one was dumped on the garbage heap. (*The Locked Room,* p. 68)

The so-called "welfare state" is cast as an environment that has nurtured the growth of unscrupulous profiteers. The murder victim often turns out to be one of these unscrupulous profiteers, as in the case of *Murder at the Savoy,* where the multimillionaire businessman, Viktor Palmgren, is murdered by an ex-employee whose life has been ruined by Palmgren's ruthless business practices. As is also the case in a number of other books in the series, the plot is crafted in such a way that the reader comes to believe that the victim, who symbolizes all the ills of the capitalist system, truly deserved to be murdered, and the reader is moved to sympathize with the murderer. Several times in *Murder at the Savoy,* we are reminded of how Palmgren ruined the lives of numerous other innocents as well. Just in case we don't see the class implications of this scenario, the authors offer a helping hand in a conversation between Beck and Inspector Per Mansson: " 'When [the murderer] began to realize that it wasn't just his own hard luck, but that he was being treated unjustly by one man or perhaps by a social group . . .' And Palmgren represented just that social group" (*Murder at the Savoy,* p. 193).

Let there be no doubt. Here again iconography and dialog combine to produce a particularly pointed analysis of the cause of the crime. What is interesting and poignant is that this interpretation, which attributes to the killer the ability to comprehend the relationship between the individual and class, is being offered by two policemen for whom the reader has gained respect over the course of the novel.

Another summary interpretation is added soon thereafter that removes any modicum of sympathy for the alleged victim, as the authors summarize the information revealed to Beck and Mansson, information that confirms their own earlier interpretations: "Viktor Palmgren, the bloodsucker, who lined his purse at the expense of other human beings, the big shot, who didn't give a damn about the welfare of his employees or tenants" (*Murder at the Savoy,* p. 198).

So why should we care about Viktor? Of course, we shouldn't. We should care about the killer who, in fact, has become the true victim in the case. Even the usually apolitical Martin Beck has been persuaded to look upon the murderer as a man wronged and victimized by the bloodsucking capitalist. A similar opinion is offered by one of Beck's colleagues.

> "I'll tell you something I've never said to anyone else," Gunvald Larsson confided. "I feel sorry for nearly everyone we meet in this job. They're a lot of scum who wish they'd never been born. It's not their fault that everything goes to hell and they don't understand why. It's types like this one who wreck their lives. Smug swine who think only of their money and their houses and their families and their so-called status. Who think they can order others about merely because they happen to be better off." (*The Laughing Policeman,* p. 210)

The authors are often compelled to give the story one final twist —to add hopelessness and futility to the mix as well. It might have been a noble and heroic act against the evil forces of capitalism, but it was most likely singular and futile, with no significant impact on the overall system.

> Viktor Palmgren was dead.
>
> Gone forever and missed by no one, save for a handful of international swindlers and representatives of suspect regimes in countries far away. They would soon learn to do business with [Viktor's successor] instead, and so things would be, to all intents and purposes, unchanged. (*Murder at the Savoy,* pp. 203–204)

It might be argued that Wahloo and Sjowall chose the police procedural genre to promote their ideological agenda because the police represent and protect the basic structures and institutions of the political-economic system and because of the close relationship between individuals and society that one finds in these novels. Within this context, and as an extension of their sympathetic perspective on killers of certain kinds of individuals, we find discussions relating to the alienation of the police from the society they are employed to serve and protect: " 'There's a latent hatred of police in all classes of society,' Melander said. 'And it only needs an impulse to trigger it off' " (*The Laughing Policeman,* p. 101).

Moving away from more ideological considerations, both philosophically and spatially, the Martin Beck series does allow readers the opportunity to visit various other regions of Sweden and Scandinavia. Murder investigations in smaller towns like Malmo and Motala provide the authors the opportunities to share impressions of regions outside of Stockholm. Generally these smaller towns are described in straightforward, matter-of-fact, journalistic style. Reference to regional accents help to educate readers to the fact that regional variations in Swedish identify different parts of the country, with a touch of ethnolinguistic centrism thrown in for good measure. At the same time, however, we are reminded of the commonalities of language throughout distant regions of Europe as, for example, in the relationship between Finnish and Hungarian: "Then he began to speak and at first Martin Beck thought the man was speaking Finnish, but then remembered that Finnish and Hungarian stemmed from the same linguistic stock" (*The Man Who Went Up in Smoke,* p. 50).

This revelation of the commonalities between parts of Scandinavia and parts of Eastern Europe occurs during a case that takes Martin to Hungary to investigate a missing Swedish journalist. It provides one of our few exposures to Eastern Europe during the height of the cold war, and descriptions of Budapest alert us to its status as one of the world's most beautiful cities.

> The Danube was flowing past him on its calm, even course from north to south, not especially blue, but wide and majestic and indubitably very beautiful. On the other side of the river rose two softly curved hills crowned by a monument and a walled fortress. Houses clambered only hesitantly along the sides of hills, but farther away were other hills strewn with villas. That was the famous Buda side, then, and there you were very close to the heart of central European culture. (*The Man Who Went Up in Smoke,* p. 39)

The success of this unequivocally ideological approach to the police procedural, as always, rests on the appeal of the main character, Martin Beck. Beck remains predominantly apolitical throughout, giving us a hard-working civil servant whose rocky marriage dissolves during the early novels in the series and who sees his job as noble and important. Yet he is a modest man, in both appearance and intelligence, a point brought home early on by the authors in a sublimely understated description of our main character.

> Martin Beck wasn't chief of the Homicide Squad and had no such ambitions. Sometimes he doubted if he would ever make superintendent . . . There were people who thought he was the country's most capable examining officer.
>
> Some women would say he was good looking but most of them see him as quite ordinary. (*Roseanna,* pp. 10–11)

To a great degree, it is because Martin Beck is capable and ordinary and because he is part of the system being critiqued that the political and ideological agenda of Wahloo and Sjowall is so provocative and compelling. It is also an excellent example of the effectiveness of developing a character over the course of an entire series. Increasingly throughout the series, Martin Beck becomes worn down, loosing any illusions he had about society and his role in it. He becomes even more human and less heroic over time, strengthening the bonds between reader and protagonist. ■■■■■■■■■■■■■■■■■■■■■■■■■■■■■■■■

Conclusion

The European continent gives ample opportunity to extend our impressions of popular, escapist literature in diverse directions. Heavy doses of social commentary add to the flavor of these offerings. Whether true native or adopted son, these four authors take approaches that are decidedly different from their American and British forebears. Questions of legality, morality, and ethics, so black and white in much of the traditional police procedural literature, seem blurred and ambiguous. The questioning of the system within which the police, as extensions of social stability, operate raises critical societal questions. These societal questions are closely tied to the places where they evolved, and they have grown into reflections of these similar, yet strikingly different locales for murder.

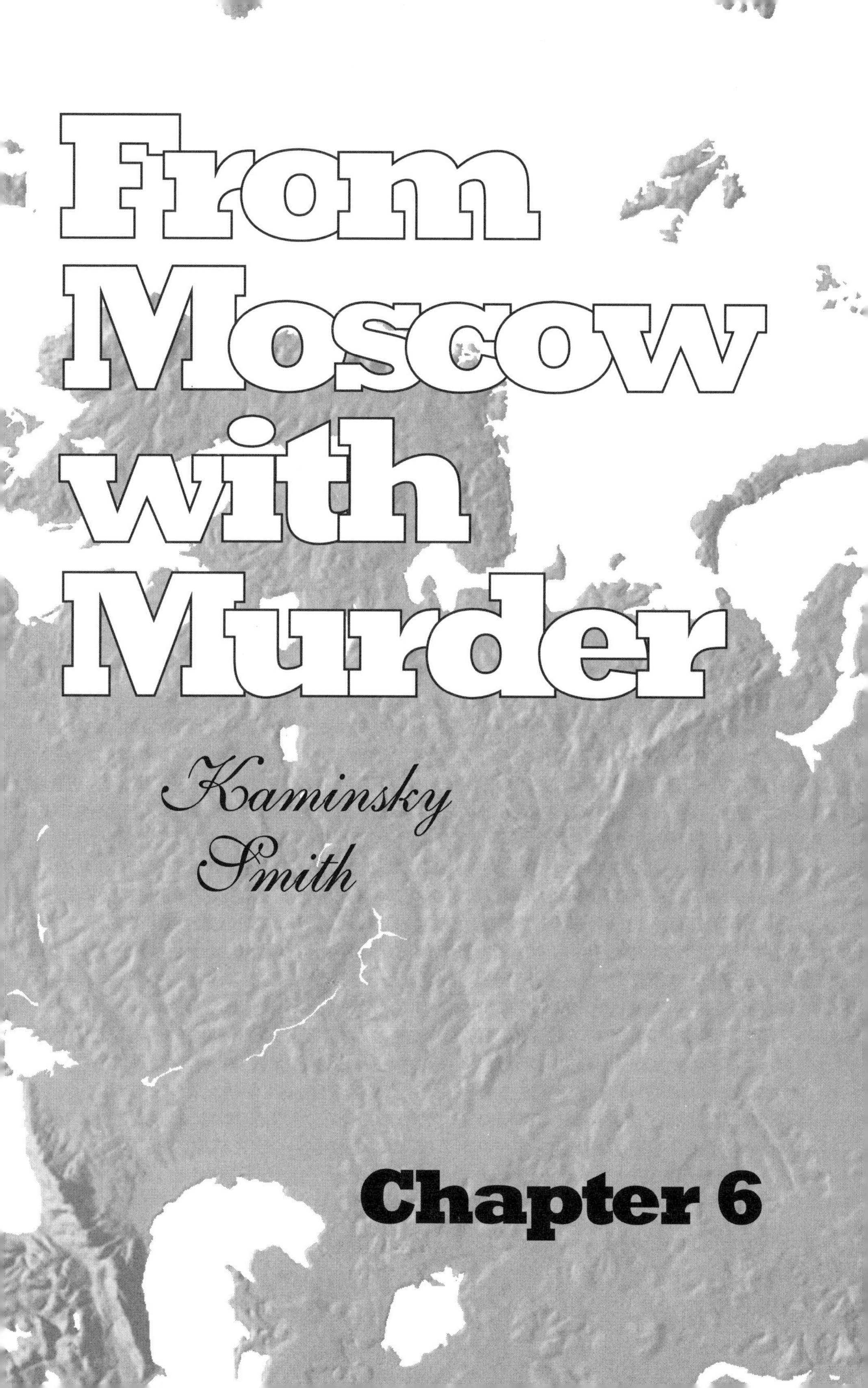

From Moscow with Murder

Kaminsky

Smith

Chapter 6

The cold war socialist world is greatly underrepresented in the police procedural genre of literature. Mysteries concerning the countries of Eastern Europe and the Soviet Union are predominantly spy novels, which produce a rich and popular literature that often has a strong sense of place that concentrates on stereotyping the extremes of socialist society—either highly romanticized or overly paranoid. The everyday, mundane world of the cop on the beat is not part of this world, which was particularly true for Russian writers during the Soviet period. Detective fiction was not an acceptable genre because it necessitated commenting upon the ills of society, an exercise that would never have passed state censors. In the early 1980s, the first incursions into the world of the Soviet police were written "from the outside, looking in," with the introduction of two serials based on Moscow detectives: Martin Cruz Smith's Arkady Renko and Stuart Kaminsky's Porfiry Rostnikov. In addition to being very popular with the public, both writers received awards for books in the series—Smith's *Gorky Park* received the Golden Dagger Award and Kaminsky's *A Cold Red Sunrise* received the Edgar Award. Both series have been best-sellers.

For those interested in things Russian, the series uncovered a Soviet society that was far more complex than many Westerners had imagined. Black-and-white images were transformed into numerous shades of gray in the minds of many readers as they were exposed to the details of life in Moscow and the intricacies of Soviet society. For those interested in mysteries, both series possessed all the elements of a good murder novel—character development, an intricate, yet believable plot, and a flow that kept the reader involved from the start. ■■■■■■■■■■

Martin Cruz Smith

"You didn't quit the [Communist] Party,
you were thrown out, that's why I trust you . . ."
(Red Square, p. 45)

MARTIN CRUZ SMITH was the first to provide Westerners with a master Moscow detective—Chief Inspector Arkady Renko, who has served as the central character for four murder mysteries. The first, *Gorky Park,* takes place during the last years of Brezhnev, the period of quintessential cold war Soviet Russia for many of us. As punishment for irregularities in his investigation of the three dead bodies in Gorky Park, Arkady is banished to slime fish on a factory ship in the Bering Sea in *Polar Star,* where his success in solving the murder of a female crew member gets him reinstated in the militia. Arkady returns to Moscow in *Red Square* to solve an apparent mafia murder during the Gorbachev era. And an Arkady of the post-Soviet period appears in 1999 in *Havana Bay.* Unfortunately for those interested in better understanding Moscow, the novel takes place entirely in Cuba. Although post-Soviet tension between Russia and Cuba forms a recurrent theme, only intermittent flashbacks reveal the decay that has consumed life in present-day Moscow.

For many Westerners the characters Smith created present an insightful glimpse into a society far more diverse and complex than many had believed. This was particularly compelling for the Soviet Union during the late Brezhnev era because the very personal accounts of murder are juxtaposed with a constant barrage of stereotypes, reinforcing the general impressions we had of a fairly homogenized society, tightly and effectively controlled by a monolithic government. Because *Gorky Park* attacks these stereotypes in a very personal and individual way, it is particularly imperious. This contributes to its ability to convey sense of place so effectively. It may be fair to assume that whenever a reader is not concentrating on the possible identity of the murderer, there remains great fascination with events taking place in Moscow. This makes the reader even more open to the suggestions that comprise the sense of place.

Smith is a master of narrative description as the foundation for "seeing" locales crucial to the plot. In these narratives, he describes both setting and mood.

> Why Gorky Park? The city had bigger parks to leave bodies in—Izmailovo, Dzerzhinsky, Sokolniki . . . It was the first park of the Revolution, though, the favorite park . . . It was the place everyone came to: clerks to eat lunch, grandmothers with babies, boys with girls. (*Gorky Park,* pp. 9–10)

Smith provides a trove of images from the start. A sense of transition moves us from the long, dreary winter to a bright, warmer spring that tries to dissipate Moscow's collective depression. Given a picture of Gorky Park, we begin to understand how important the place is to Muscovites; it is no ordinary tourist trap. It is a place for residents to escape, meet friends, and have fun. It captures the essence of all Moscow parks and their importance to Muscovites. How can someone dare to commit such a barbarism here?

Images build of a despondent society searching for direction, for meaning. Clearly, it is not easy, nor possible, nor necessary to entirely separate images, moods, and symbolism. They are often interrelated.

This picture is added to and built up throughout the novel with numerous additional, short passages that provide an ever-increasing description of the place. By the end of the novel, the reader is bundled up with a literary patchwork quilt of mood and place descriptions. More often than not, the novel does not devote large chunks to detailed descriptions of the setting. They come as small snacks, often served unexpectedly, throughout the novel.

Once we understand the environment where the plot takes place, the novel serves as the medium for understanding the society and its people. Again, the intriguing aspect of *Gorky Park* is that this component of place breaks many of the stereotypes Westerners had about Soviet society in the years before the ascension of Gorbachev. The relationship between Moscow, as center of the empire, and the periphery is subtly exposed when discussing the plight of the militia, or police.

> The militia enlisted farm boys right out of the Army, seducing them with the incredible promise of living in Moscow, that residence denied even to nuclear scientists. Fantastic! As a result,

> Muscovites regarded the militia as some sort of occupying army . . . Militiamen came to see their co-citizens as decadent, depraved and probably Jewish. Still, no one ever returned to the farm. (*Gorky Park,* p. 9)

In addressing the issue of the relationship between the capital and the provinces, two aspects of life in Russia emerge: the attraction of Moscow for all Soviet citizens and suggestions of racism. Although not quite so obvious, we can also find a hint of the labor problem endemic to Moscow, where there is a need for semiskilled workers, such as bus drivers, construction workers, and policemen, and a glut of highly skilled laborers such as scientists. The suggestions of racism are amplified later in the novel and gain their effectiveness because they are raised in scenes that are totally unrelated to the murder or the main plot of the novel; because it is not germane, the suggestion of Russian racism is simply accepted by the reader as is.

If it is true that "God is in the details," then the Smith novels have used this approach to make the description even more convincing. The mundane aspects of daily life, even if briefly mentioned, draw the reader into the feeling of being there. Little things, although they do not affect the story line, add to the credibility of the storyteller. In describing the vodka store line, the details about the return bottles provides the touch of detail.

> Drunks sagged and leaned like broken pickets on a fence . . . they clutched empty bottles in the solemn knowledge that no new bottle would cross the counter except in exchange for an empty. Also, it had to be the right size empty bottle: not too big, not too small. (*Red Square,* p. 47)

Among the strengths of a mystery series is that an author can build on earlier work. Although it is not necessary to have read *Gorky Park* or *Polar Star* to appreciate and enjoy *Red Square,* it doesn't hurt. The monumental changes that have occurred in Soviet society while Arkady was "out of town" give Smith an opportunity to comment on these changes from the character's perspective, as he does early on in *Red Square.*

> Sometimes Arkady had the feeling that while he had been away, God had lifted Moscow and turned it upside down. It was a

> nether-Moscow he had returned to, no longer under the gray hand of the Party. The wall map showed a different, far more colorful city painted with grease pencils. (*Red Square,* p. 31)

Every level of sense of place is crammed into this passage. City mood is captured in metaphors. With a lapse of a decade since the first book, Smith can convey through the eyes of a Muscovite a spirit quite different from the sense of Moscow found in *Gorky Park.* The use of metaphor is particularly effective: the turning upside down of the city, the city as nether land, the grayness of the party compared to the new colorfulness. For social interaction we find the suggestion of the declining role of the party. If we were to read on a bit, we would also see this declining role juxtaposed with the growing power of the Mafia.

Although it is true that Arkady is well aware of the radical changes that have occurred during his absence from Moscow, *Red Square* subtly suggests that much remains the same. The black market may provide a wide range of goods and services, but state stores, where the average Muscovite still shops, remain poorly stocked. But of course, in the back cooler, with the right connections, are foodstuffs readied for delivery.

> Arkady opened a bin to see sacks of flour stacked like sandbags. When he opened another, pomegranates rolled around his feet and over the storeroom floor. A third bin, and lemons poured over pomegranates . . . The last bin was stacked with cigarettes. (*Red Square,* pp. 24–25)

Parts of the old system persevered. *Pod stolom,* under the table, proves as important during *perestroyka* as it had been before. The same can be said for the Communist Party. Although the party has been discredited, it is still a powerful force in Gorbachev's Soviet Union: "'But they're Party members. Even if they quit the Party, even if they become a people's deputy, in their hearts they're Party members.'" (*Red Square,* p. 45).

Arkady Renko is the quintessential Muscovite—the average, hard-working Russian, struggling within and trying to survive in a complex system that does not always seem fair. What is surprising to many Westerners is that in the end, given the opportunity, Arkady is not looking "to escape to the West." He is Russian. He is a Muscovite. Moscow, Russia, may not be great, but it is home.

Irina, on the other hand, is emblematic of that segment of Soviet society ready to embrace the West. In Russia, she is frustrated and seeks the freedom and opportunities of the West. In the interplay between Arkady and Irina are reflected the antagonisms of Soviet society.

Irina's perspective on life in Soviet society is captured eloquently in the metaphor she uses to describe the Siberian dilemma, which Smith, in turn, uses as a metaphor for the hopelessness and helplessness of life in the Soviet Union.

> "You know what the 'Siberian dilemma' is?"
>
> "No."
>
> "It's a choice between two ways of freezing. We were out on a lake fishing through the ice when a teacher of ours fell through. He didn't go far, just down to his neck, but we knew what was happening. If he stayed in the water he would freeze to death in thirty or forty seconds. If he got out he would freeze to death at once—he would be ice, actually . . . He looked up at us; I'll never forget that look. He couldn't have been in the water for more than five seconds when he pulled himself out."
>
> "And?"
>
> "He was dead before he stood up. But he got out, that was the important thing. He didn't just wait to die." (*Gorky Park,* p. 253)

It is hard to imagine a clearer explanation of why Soviet society was willing to put up with so many lies and indiscretions on the part of its government than in Irina's explanation for her unswerving faith in Osborne's ability to get her out of the Soviet Union: "'It doesn't matter how ridiculous a lie is if the lie is your only chance of escape,' she said. 'It doesn't matter how obvious the truth is if the truth is that you'll never escape'" (*Gorky Park,* p. 260).

Arkady Renko gives insight into a complex, often chaotic society at a time when Westerners saw it as monolithic, ordered, and stable. After reading the first three novels of this series, one is inclined to ask if there were any question that the Soviet state would crumble of its own weight. Now the series serves as a critique of the past, and we can only await an Arkady Renko of the post-Soviet period. ■■■■■■■■■■■■

Stuart Kaminsky

[Rostnikov] was a Russian, a Muscovite.
It wasn't just a matter of love or loyalty. It was part of him.
(Death of a Dissident, P. 96)

STUART KAMINSKY is a prolific writer of Moscow murder mysteries. His master detective, Porfiry Rostnikov, in many ways is an older version of Arkady Renko. The first novels of each series—*Gorky Park* and *Death of a Dissident*—were both published in 1981. Although each Porfiry Rostnikov novel is about half as long as the Renko novels, the series now exceeds ten volumes. The exploits of Rostnikov span a decade that covers the late-Brezhnev, Gorbachev, and Yeltsin years, including post-Soviet Russia. Among the endearing qualities of this series are protagonists who remain much the same throughout the entire series. As with the Smith series, it is not necessary to have read earlier novels, but it does help, especially in building upon the relationship between the reader and the characters.

Kaminsky does not devote nearly as much time to outright description as does Smith, but he employs the highly effective technique, with a few exceptions, of feeding the reader bits and pieces throughout, rather than feeding large doses of description all at once.

Initial introduction to the world of Porfiry Rostnikov begins with a comparative description of setting and slips immediately into an analysis of the Russian psyche.

> Moscow winters are really no worse nor much longer than the winters of Chicago or New York. If they seem so, it is because Muscovites like to think of their winters as particularly furious. It has become a matter of pride, an expression of unnecessary stoicism somewhat peculiar to the Russian psyche. (*Death of a Dissident,* p. 1)

Kaminsky often uses a comparative approach, both to help the reader better understand and relate to the story and to put Russia and Russians in context. A little later on in *Death of a Dissident,* he again relies on the comparative approach to describe night life in Moscow. Because it is so different from night scenes in major Western cities, the comparison with New York and Paris accentuates the lack of activity, the desertedness of the cultural center of socialism. Even readers who have never been to New York or Paris are well aware of the twenty-four hour a day lifestyles of these cities and therefore can use them as points of reference.

For the most part, however, Kaminsky uses straightforward description, painting a picture of Moscow not dissimilar to the picture painted by Smith: lots of gray, lots of drabness, lots of "gigantism," lots of lifelessness in the inhabitants. Like Smith, Kaminsky is sensitive to interjecting a sense of history into setting, creating a more complete sense of place.

> Not far from the Kremlin is one of the busiest intersections in Moscow, Dzerzhinsky Square, where as many as half a million people come each day . . . most come to two massive buildings. One building is the Detsky Mir or Children's World, the biggest children's store in Russia. The other building is a strange, hulking creature in two sections . . . One half of the building pre-dates the Revolution. The other half was completed in 1948, using the labor of captured German soldiers . . . the German soldiers were reportedly executed so that they could not divulge information about the labyrinth of rooms they had built. The building . . . [is] the Lubyanka, which houses the K.G.B. (*Death of a Dissident*, pp. 55–56)

The complex nature of one Moscow intersection combines physical appearance of the buildings with the human interaction of high daily use by Muscovites. How rich an irony that the country's largest children's store is catty-corner from the center of state oppression in Lubyanka. Kaminsky ties in a sense of history by describing how the latter addition of the complex was constructed and throws in the irony of having to execute the builders for security reasons, not unlike the fate of the architect of St. Basil's Cathedral. This is all by way of prelude to Rostnikov's first encounter with the KGB.

In *Red Chameleon* Kaminsky describes the history and geography of Moscow to provide a backdrop to the plot, as only one element in uncovering the identity of a sniper loose in Moscow, a description interjected between discussions of the main murder Rostnikov was attempting to solve, the sniper Karpo was attempting to locate, and Rostnikov's concern over Karpo's injured arm. Yet the basic layout of Moscow is exposed very effectively. Moscow, in bits and pieces, is an ongoing presence throughout most of the novels, and the map of Moscow appears as an essential clue in the detection of some of the crimes.

The first of three diversions from Moscow in this series is to Siberia, where Rostnikov and Karpo are sent because of the importance of the crime to central authorities. A first introduction to Siberia involves Rostnikov's attitude toward this most mysterious of Russian places, an attitude not unique among Muscovites: "Rostnikov had never been to Siberia. He had no curiosity about Siberia. He did not want to go to Siberia" (*A Cold Red Sunrise,* p. 25).

Once there, Rostnikov proceeds as he would with any murder. Yet, before he gets started, Kaminsky feels compelled to put Siberia into Russian context. He provides a detailed history of the conquest of Siberia for four full pages (*A Cold Red Sunrise,* pp. 56–59). This varies from his more standard bits-and-pieces approach. The setting is thus provided for better understanding the basic antagonisms between Mother Russia and the Siberian periphery as suggested throughout the remainder of the novel.

Rostnikov's second venture outside Moscow occurs in *Death of a Russian Priest,* which takes place after the fall of the Soviet Union. The primary image portrayed is the contrast between the rural and the urban, an important consideration throughout the entire history of Russia and the Soviet Union. Overlaid are the antagonisms of church and party and the clash of values between rural Russian and urban communist. Rostnikov's arrival in Arkush sets the tone for the novel: "They passed small ancient houses of wood and stone along the cobbled street. It struck Rostnikov that he had gotten off the train and stepped into the past" (*Death of a Russian Priest,* p. 66).

Kaminsky is also an effective conveyor of details whereby he establishes his credibility, verifies the authenticity of the description, and secures the trust of the reader.

> Left-handedness was discouraged in Russia. Russian children caught using their left hand to throw, write, or eat were sternly stopped. Karpo had never thought much about this, assuming the idea of conformity was simply part of one's education in an overpopulated society. (*Red Chameleon,* p. 27)

The addition of this little bit of insight into Russian culture is so striking because it does not need to be there. It adds nothing to the main plot. It isn't even essential to Karpo's struggle over what to do about his injured arm. Yet it is precisely this kind of attention to detail that tells the reader that Kaminsky knows what he's talking about. His familiarity with the importance of being right-handed in Russian society confirms his expertise and ability to tell the story. It helps to open or keep open the mind of the reader to accept the author's vision of reality.

One of the defining features of Moscow street scenes is its infamous lines. What description of Moscow would be complete without at least one encounter with a line? In the details of the line is credibility gained, with the reader coming away with a feeling that "the place is familiar."

Kaminsky does not simply provide a description of a line; rather, he delves into the entire laborious process of shopping in Russia, a process that includes a series of lines and divides the shopping experience into tedious and seemingly needless encounters with the various controllers of the goods being sold. And as if the description of process weren't enough, Kaminsky provides a sermon on the importance of lines to Russian society.

> Lines are a way of life in Russia . . . People go mad in lines, come to the major decisions of their lives in lines, get their principal education and entertainment from the books they read in lines, and make lifelong friends and enemies in lines. (*Death of a Russian Priest,* p. 32)

Just as easily, however, the lack of attention to detail can foil the entire process of building credibility. In *Blood and Rubles,* Kaminsky damages years of credibility and plausibility by making a serious error in the initial presentation of three concurrent crimes, all of which supposedly occur in Moscow. In one case, unfortunately, he clearly describes a scene from St. Petersburg, putting the crime "[a] few short blocks from the Neva River, not far from Saint Isaac's Square . . . Elena had taken the number 3 bus down Nevsky Prospekt . . . [and] Karpo . . . had arrived by metro at the Gostinniy Dvor stop" (*Blood and Rubles,* pp. 8–9). This is unmistakably St. Petersburg. As the story unfolds, there is no question that it was supposed to be in Moscow. A major mistake like this in one of the earlier volumes could have killed the series at the start. Occurring in the tenth novel, after years of building a loyal following, it may be forgiven, especially because it does not affect the resolution of the crime, after the initial confusion about why Elena and Karpo are investigating a crime outside of Moscow is cleared up, nor significantly detract from the overall plot of the story.

It is almost a truism that contradictions are found in all societies. Yet they are particularly important in novels about the Soviet Union because they were at the very basis of Russian society during the Soviet period—incongruities between people's public and private lives, what they said in public and what they said in private, how they paid homage to communism in public, while privately holding on to their beliefs in orthodoxy, and the list goes on. Contradictions were not simply a part of Soviet society, they were part of its foundation.

> [I]nquiry is based on the frequently stated assumption that "every person who commits a crime is punished justly, and not a single innocent person subjected to criminal proceedings is convicted." This is repeated so frequently by judges, procurators, and police that almost everyone in Moscow is sure it cannot be true. (*Death of a Dissident,* p. 10)

Each case is symbolic of the deeper inconsistencies that pervade Soviet society, not only in the relationships between people, but also in the relationship between the people and their leaders. Nowhere is this more acute than in the public display of atheism and the private belief in God, a point often, if only briefly touched upon throughout the novels: "Boris led the mother and boy toward his [ice cream] stand, praying to the gods that didn't exist that he would never see the pale man [referring to Karpo] again. And the gods that didn't exist granted the wish" (*A Cold Red Sunrise,* p. 31).

In the Kaminsky novels, it is difficult to separate human interaction from iconography because the primary strength of the Rostnikov series is the use of the main characters as icons of certain segments of Russian society. Rostnikov is an older, married with child, version of Arkady Renko. His wartime record, which has left him with an injured leg that is a constant burden, proves his patriotism; yet, he is a pragmatic person, not ideologically driven: "Rostnikov had difficulty accepting the priorities of his society. He recognized them, understood them, sympathized with them, but it was difficult. He had perhaps read too little Lenin and too much Dostoyevsky" (*Death of a Dissident,* p. 120).

Whereas Arkady's marriage falls apart during the early stages of *Gorky Park,* Porfiry has been married for many years and has a son. Family interaction is constantly interjected into the plot. It complicates matters because his wife, Sarah, is Jewish and wishes to leave the Soviet Union, and his son, Iosef, is serving his military duty in the army, which sends him to Afghanistan.

> Sarah . . . belonged to an official national group in Russia. The Jews had to register just as did the Armenians or any other ethnic segment of the populace. Sarah was so registered. They had married after Porfiry was a policeman. If they had married before, it was almost certain that he would never have been given the job he was so good at. In fact, it was only his reputation that protected him. (*Death of a Dissident,* p. 94)

Kaminsky uses Sarah to comment on the plight of Jews in Soviet society, a theme that winds its way throughout the novels. Initially, the possibility of leaving the Soviet Union is introduced. In succeeding novels, the Rostnikovs try unsuccessfully to emigrate, and then they must deal with the consequences of remaining in the Soviet Union after trying to get out. Porfiry is demoted but is too valuable to lose his job entirely, as happened to many Jews who unsuccessfully tried to emigrate. But the failure had a more dramatic effect on Sarah, who lost her job and became physically and psychologically ill in the aftermath.

Kaminsky keeps two police colleagues, Sasha Tkach and Emil Karpo, in addition to Porfiry's family, constant throughout the series. Tkach is a police officer in his late twenties. As the series begins, he has recently married. In addition to his exploits as a young policeman who gains experience and expertise under the tutelage of Rostnikov, his family life provides insights into the typical struggles of an extended and growing Moscow family.

Possibly the most intriguing of all of Rostnikov's colleagues, however, is Emil Karpo, referred to as the "Tartar" or "Vampire" for his rather foreboding appearance. He embodies all the qualities of *homo Soveticus,* the new Soviet man. He is totally dedicated to his work, which he puts into the larger context of protecting the goals of the revolution and the sanctity of the Soviet state.

Unlike others who are ambivalent about communism and the Soviet system, Karpo is a true believer. When someone tells Karpo that he knows his Marx, acknowledging how well he was versed in communist ideology, Karpo replies: "I believe my Marx." If ever there were a counterbalance to Smith's Irina, it is Kaminsky's Emil Karpo.

> To Karpo, Russia meant sacrifice. The revolution was far from over, might never be over. There was only the struggle, the dedication, the small part one could play in the bigger picture. There wasn't necessarily a victory to be achieved. Life was a series of tests, challenges that one was either prepared for or would be worn away by. Since hardship was inevitable, it was best to condition oneself to it. (*A Fine Red Rain,* p. 14)

And if there were any doubts of Karpo's sincerity, they are forever erased when he is faced with the possibility of dying in the line of duty: "'It is my duty,' said Karpo. 'If it costs me my life, then I will die in the course of my duty. If I value my life more highly than I value the meaning by which I live, then my life has no meaning'" (*Rostnikov's Vacation,* p. 124).

His undying devotion to communism makes for an intriguing personal struggle after the collapse of the Soviet Union in *Death of a Russian Priest.* He must now reconcile the loss of the communist system that had guided his life and his dedication to his job and duty as a policeman. You do not have to be a communist to empathize with the doubt, confusion, and anguish that Karpo, as a symbol of all true believers, experiences.

> The union was gone. The Soviet Socialist Republics were now a commonwealth of sovereign states. Leningrad was once again St. Petersburg. They had even gotten rid of the hammer and sickle and designed a flag that seemed no flag . . . Meaning was disappearing, but what little there was he clung to. His faith and loyalty had lost their certainty and there were moments when panic threatened to break through. (*Death of a Russian Priest,* p. 50)

Emil Karpo is still struggling with this dilemma two novels later in *Blood and Rubles.* Kaminsky titillates the reader with unfinished business in each of his novels, leaving us to anticipate the next episode. We still anticipate how Karpo will work out his internal struggles, whether the Rostnikovs may once again attempt to emigrate, and we look forward to the arrival of the second Tkach child. But all this must wait for the succeeding novels. Such is the lure of the well-done police procedural series. ■■■■■■■■■■■■■■■■■■■■■■■■■■■■■■■

Conclusion

These two series paint a picture of a very complex and unstable society in the Soviet Union during its final decade of existence. They represent our most compelling examples of detective fiction there. After the fall of the Soviet Union, numerous stories of murder use Russia as setting; more dramatic, however, Russian writers themselves are producing mysteries en masse, and they sell, a rewarding legacy for Smith and Kaminsky, who still stand as the primary conveyors of sense of Russia to Western readers during the final stages of Soviet control.

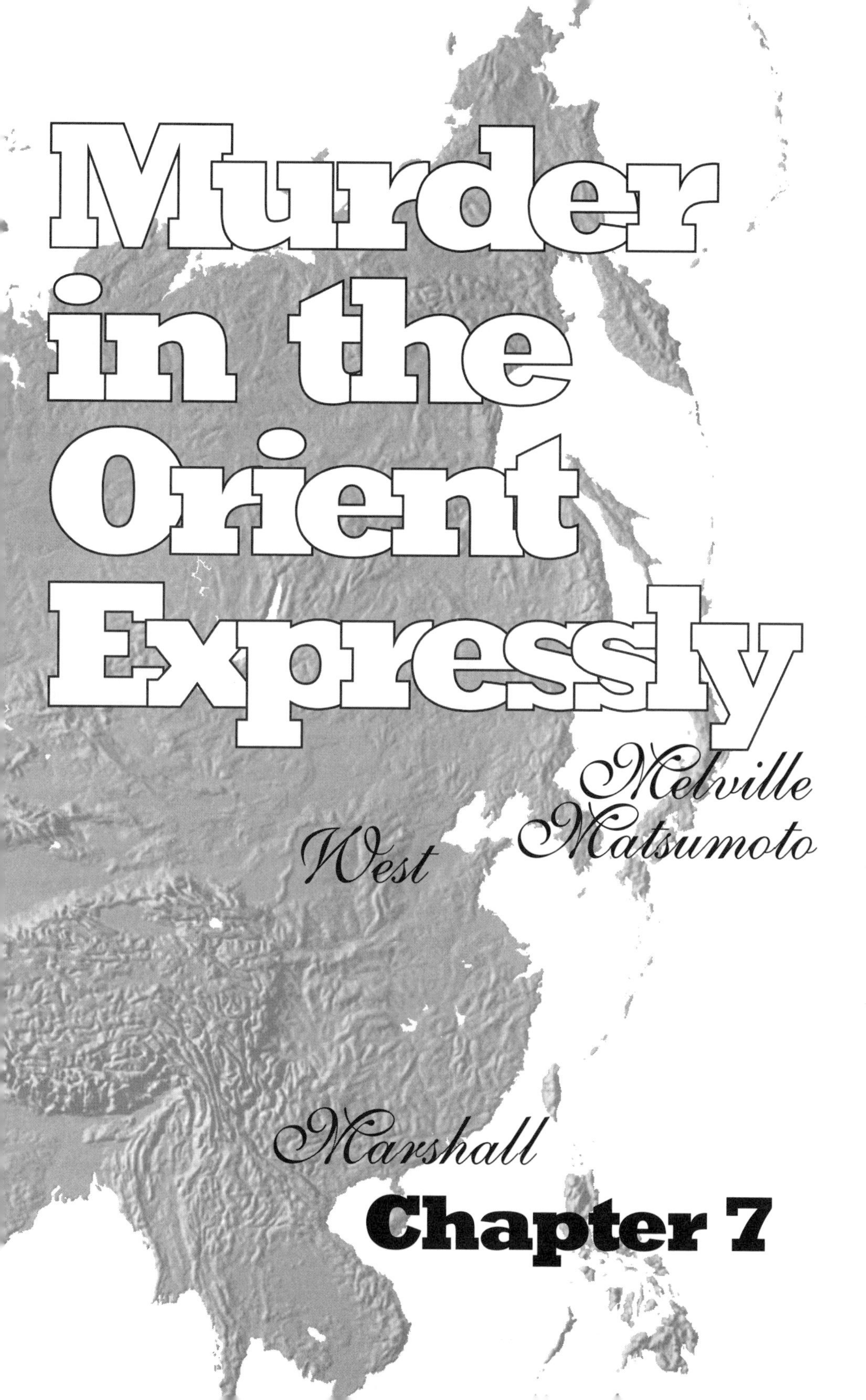

Murder in the Orient Expressly

Chapter 7

For centuries Westerners have been enthralled by the mysteries of the Orient. The authors presented here unravel these mysteries in quite diverse ways. For Japan, I compare and contrast two authors, one native and one foreign, juxtaposing an insider's view with an outsider's view. Seicho Matsumoto offers an insider's view. Although his is not a single series, translations of two of his police procedurals provide clues to a Japanese interpretation of the police procedural genre. James Melville offers an outsider's view with his Kobe-based series featuring Police Superintendent Tetsuo Otani. From Japan we travel to Hong Kong. William Marshall's Harry Feiffer and the Yellowthread Street crew expose us to a comedic, at times farcical, look at this last outpost of the British Empire. Our visit to the Orient concludes in Beijing, where a singular effort by Christopher West and his Inspector Wang Anzhuang reveal a complex communist context for murder. This first novel has the potential to become a successful series, and it is included here because it deals with a place about which there is little escapist literature. What little escapist literature there is generally involves espionage and international intrigue, but nothing about the day-to-day workings of life in communist China. It reminds one of the early 1980s when Martin Cruz Smith and Stuart Kaminsky gave readers their first glimpses of communist Russia. And as a brief addendum—good news! As we go to press, a second novel in the Christopher West series has appeared. ■■■■■■■■■■■■■■■■■■■■■■■■■■■■■■

The Japan of Seicho Matsumoto

There was *another region where the Tohoku dialect was spoken. Kuwahara had found another source for Imanishi. This was called* Map of Japanese Dialects.
(Inspector Imanishi Investigates, P. 98)

SEICHO MATSUMOTO WAS a prolific and noted writer of mystery novels and short stories, several of which are police procedurals. Most of his novels and short stories were written originally in the 1950s and 60s, and only in the late 1980s and 90s have a number of his works been translated into English. Although they are not part of a series based on the same protagonist, two of his translated novels fit nicely into the category of place-based police procedurals, providing a distinctly Japanese approach to the genre. *Points and Lines* and *Inspector Imanishi Investigates* share several common themes: the importance of the rail system, regional dialects as indicators of cultural regionalization, and the value of maps in the search for patterns. *Points and Lines* adds a heavy dose of the cozy relationship between business and government as a major plot element. Although it is not an overriding plot element in either novel, allusions to the aftermath of World War II are sprinkled throughout.

Whereas *Inspector Imanishi Investigates* focuses on a single, Tokyo-based detective, Imanishi Eitaro, *Points and Lines* features a Tokyo-based detective, Kiichi Mihara, who is helped by a local Hakata detective, Jutaro Torigai, playing off the young, sophisticated urban cop against his older, small-town counterpart.

The resolution of the crime in *Points and Lines* depends almost entirely on Inspector Mihara's understanding of time-space geography and the ability of modern technology to overcome "friction of distance." Even though he is convinced he knows the identity of the murderer early on in the novel, he undertakes a laborious effort to coordinate train and airline routes and schedules to confirm that the suspect actually could have been at the scene of the murder at the appropriate time.

"Hokkaido is a surprise! The other end of Japan!"

"Yes, quite a distance from Kyushu. Disappointing!" Mihara said with feeling. (*Points and Lines*, p. 83)

Just in case we're not quite sure how disappointing this is to Inspector Mihara, a map of Japan reinforces that these two places are about as far away from each other as one can get in Japan: "'. . . from the extreme west to the extreme north!'" (*Points and Lines,* p. 82).

Once Mihara resolves an apparently airtight alibi, based on the premise that you can't be in two places at the same time, the case can be concluded successfully. In the meantime, Mihara travels the extent of Japan as part of his investigations. As we travel with him, several additional characteristics emerge. One is the great variety of dialects, revealing a more culturally diverse society than most Westerners are cognizant of. The general picture of the Japanese as a homogeneous people, united by ancestry, language, culture, religion, and historical background, is, for the most part, correct. Yet, there are cultural differences, marked first and foremost by dialects. In Hakata, in northern Kyushu, the investigation focuses on a couple that could very well have been the murderer and his victim. When asked about the couple, a local resident has no doubt that the possible victim-to-be was not a local: "'No, it was a pleasant voice. But she didn't have the local accent. People around here don't speak like that. I believe it was a Tokyo accent'" (*Points and Lines,* p. 44).

Attention to dialects and regions of origin plays an even more important role in *Inspector Imanishi Investigates,* where Imanishi is hard pressed to learn where the victim, found without identification on the tracks of a Tokyo train station, is from so that he can identify him. It all centers around the dialect of the victim, who was overheard speaking prior to his death. Yet lack of knowledge of various dialects leads Imanishi on a wild goose chase to the Kameda region prior to figuring out that the region of dialect was actually Kamedake: "According to Imanishi's research at the language center, the people of this area swallowed the ends of words. What the witnesses had heard as Kameda had actually been Kamedake" (*Inspector Imanishi Investigates,* p. 100).

Combining travel with investigations unveils another common thread in the Matsumoto series—the value of local knowledge and the importance of a sense of place. Appreciation of place takes many forms, as in the case of Imanishi's colleague Yoshimura, whose sense of place revolves around taste.

> "We go on various business trips like this. And afterward, rather than the scenery or problems I might have encountered, what I remember is the food. Our expense allowance is so small we can only afford rice with curry . . . Yet the flavoring is always

different. It's the taste of each location that I remember first." (*Inspector Imanishi Investigates,* p. 20)

The importance of local knowledge goes beyond just an appreciation for regional variations of different sorts. It often unlocks clues to the investigation. In *Points and Lines,* not only did the victim's accent suggest that she was not local, her unfamiliarity with the area helps to confirm who she is: "The implication was quite clear: Kenichi Sayama had made many trips to Hakata on business; therefore he knew the locality well. Otoki, on the other hand, had not been in Hakata before" (*Points and Lines,* p. 48).

One of the great strengths of Matsumoto's mysteries, as they have been translated into English, stems from their sensitivity to spatial relationships, which are a fundamental part of all his plots. ■ ■ ■ ■ ■ ■

The Japan of James Melville

"Apart from China I don't believe there's such a totally integrated society as ours in the world."
(The Chrysanthemum Chain, P. 202)

IN CONTRAST TO Matsumoto's insider's interpretation of place, James Melville offers an outsider's view, albeit an outsider who has extensive experience in Japan. Melville spent seven years with the British Council in Japan and later returned to Tokyo for four years as head of the British Council and as cultural counselor. His love of the culture and respect for the people comes through clearly in his writings. For nearly two decades, Melville has crafted a Kobe-based police procedural around the personality of Police Superintendent Tetsuo Otani. Several themes recur in the Otani series.

The homogeneity of Japanese society is oft discussed in these novels, even though Melville introduces numerous examples of the cultural diversity of the Japanese. At the same time, the Otanis are used as barometers of the changing nature of Japanese culture and society. Tied to the changing nature of Japan are references, often just in passing, to the legacy of World War II. As in Matsumoto's novels, the key role played by the rail system appears as background to movement along the archipelago. Yet the major difference between Matsumoto and Melville's approaches to Japan is the overbearing presence of foreigners, especially the British, in all the Otani novels. They appear in great abundance as victims, suspects, and murderers. In *Death of a Daimyo,* the Otanis even manage a trip to London to visit their daughter and son-in-law and fully experience British culture.

First and foremost, however, the locale for murder is Japanese society, where even the Japanese acknowledge the homogeneous nature of their society: " 'Ever notice how often we talk about "We Japanese"? . . . We're all part of the organism, and we all think and work in social terms. If society is corrupt, we're all part of the corruption ' " (*The Chrysanthemum Chain,* p. 202).

Yet, the nature of this society is changing, and both Tetsuo and his wife, Hanae, serve as describers and critics of the process of change.

> Hanae herself had not been out in the evening even once during the whole of [their daughter] Akiko's childhood. This was the common experience of middle-class women of her generation, and though she accepted that times had changed even in Japan, the idea of leaving the three-year-old son in the care of a cheerful but unquestionably scruffy student called Rosie had appalled Hanae. (*Death of a Daimyo,* p. 26)

To further emphasize the nature of change in Japan, World War II is used as more than simply an historical dividing line between the old Japan and the new: "Hanae was a child during the war, but she's quite old enough to remember what it was like to go cold and hungry in the bitter years that followed the surrender" (*The Body Wore Brocade,* pp. 4–5).

As many changes as one finds in Japan, traditional Japanese culture perseveres in many forms, often in the form of Shintoism.

> It was the second of January . . . and the road leading up the slope to the Ikuta Shrine in central Kobe was closed to vehicles . . . to enable tens of thousands of people to make their way into the precincts and approach the sanctuary. (*The Body Wore Brocade,* p. 2)

The intrusion of foreigners into Japan is the basis for *The Wages of Zen,* when a murder occurs in a Zen study center, where foreigners come to learn about and practice the teachings of the great Zen masters. What better way to highlight the ubiquitous nature of foreign intrusion than by its penetration to the heart of traditional Japanese culture? Foreign intrusion, however, is not limited to cases investigated by Inspector Otani. He himself has been seduced by certain aspects of Western culture. He is unrepentant, for example, about his preference for a Western-style breakfast of ham, eggs, toast, and coffee over the traditional Japanese fare of rice and fermented soybean soup. Most noticeably, however, his membership in the Rotary Club is an important part of his life and often figures in the plots of his cases, although there is a decidedly Japanese flavor to the requirements of the Japanese variant.

> [H]e had to admit in his innermost heart that the iron requirement to attend the regular luncheon meeting . . . was an irksome thing. He had heard of American Rotary Clubs which, unabashed, admitted to appalling attendance rates of eighty per cent or less, and sometimes felt a little subversive about Japanese dedication. (*The Wages of Zen*, p. 2)

It appears, unfortunately, that the series may be at an end. The most recent novel, *The Body Wore Brocade*, published in 1992, employed an ingenious twist to plot development. The case is presented as an interview by James Melville of the now-retired Inspector Otani. Thus, the story line is delivered through Otani's own words. We must wait to see if Melville returns to interview the good inspector once again. ■ ■ ■ ■ ■ ■

The Hong Kong of William Marshall

The Hong Bay district of Hong Kong is fictitious,
as are the people who, for one reason or another, inhabit it.
(Epigraph to *Thin Air*)

THIS SEEMINGLY INNOCUOUS DISCLAIMER, provided as an epigraph at the beginning of each novel, sets the tone for the entire series. Described as an "audacious mixing of near-slapstick comedy with mass murder, realism with pulp-style fantasy," the Yellowthread Street series of William Marshall may defy characterization into one particular genre of popular literature. There are, however, enough basic components to include it in any discussion of police procedurals.

Yellowthread Street Station, in the Hong Bay district of Hong Kong, is the fictitious setting for this series, akin to Ed McBain's 87th Precinct and Peter Turnbull's P Division, yet with a strong comedic bent. In addition, the Yellowthread Street series has a main character throughout—Detective Chief Inspector Harry Feiffer, a third-generation Hong Kong native. Several other characters remain constant throughout the series as well—Detective Senior Inspector Christopher O'Yee, Feiffer's half-caste partner, and Detective Inspectors Phil Auden and Bill Spencer.

An intrigue blending the real and the fictional is also offered early in the story line of each novel, where a brief description of the real Hong Kong helps to introduce the fictional Hong Bay.

> *Hong Kong is an island of some thirty square miles under British administration in the South Sea facing the Kowloon and New Territories area of continental China . . . The climate is generally sub-tropical, with hot, humid summers and heavy rainfall. The population of Hong Kong . . . is in excess of five million. The New Territories are leased from the Chinese. The lease is due to expire in 1997 . . . Hong Bay is on the southern side of the island and the tourist brochures advise you not to go there after dark.* (*Thin Air,* pp. 17–18)

For the most part, the sense of Hong Kong proffered by this series is not primarily through the use of landscape description, of which there is little. The key features of the region that are often alluded to are the oppressive heat, constant noise, incessant overcrowding, and importance of water. For this series, one primary sense of place concerns racial antagonisms. Iconography becomes a key means for conveying sense of place. There are four basic kinds of characters: the European, who believes in the superiority of Europeans over Orientals; the half-caste; the Oriental; and a person of European descent who is culturally sympathetic to the Oriental. In *Thin Air,* where racial antagonisms are a fundamental component of the plot, Inspector Dobbs represents the first group, O'Yee represents the half-castes, and Feiffer is the person of European descent who is sympathetic to Orientals.

In discussing the negotiations that are taking place with the "number two" man behind the plot, a conversation between Dobbs and Feiffer gets to the heart of the problem.

> Dobbs said, 'The Chinese are stupid. Without a white man to direct them, they couldn't even tie their shoelaces on the right shoes.' He said, 'They're stupid.' He asked, more than a little surprised, 'You didn't know that? Someone told me you'd been in the Colony a long time—'
>
> Feiffer said, 'I haven't been in the Colony a long time. I was born here. I live here. I intend to go on living here and unless anything dramatic happens, like my father and my grandfather before him, I'll probably die here.' He said, 'I hope that makes my position clear to you.'
>
> There was a pause. Then Dobbs said suddenly, 'Oh—!' (*Thin Air*, p. 37)

In this particular novel, the racial antagonisms are more than simply backdrop or social commentary; they are an important part of the rationale for the crimes, which are engineered by one of Dobbs' Chinese inspectors who feels he has been slighted by Dobbs and the system because other, less-capable European cops have been promoted over him and he is constantly harassed by Dobbs. The plan to destroy the Hong Kong airport to prove the superiority of the Chinese mind over the European was motivated by these antagonisms.

> 'Mr Dobbs here, after near enough to twenty years in the Colony, has never lowered himself to learn or retain more than two words in Chinese.' He said, 'All Mr Dobbs' Chinese staff speak perfect accentless English or Mr Dobbs' Chinese staff don't get promotion . . . Munday, the Chief Inspector, hasn't half my experience or half my intelligence, but he's a Chief Inspector. But then, Mr Munday is a White Man.' (*Thin Air*, p. 177)

Although racial antagonisms are the foundation of *Thin Air*, this theme is repeated in numerous other stories in the series, continually reminding us of the diverse and complex mix of ethnic groups in present-day Hong Kong. It also reminds one of the legacy of British rule over the soon-to-be-former Crown Colony. In *The Far Away Man*, the legacy of British rule is attacked explicitly, centering on a traditional British overseas club, the Windjammer Club. It affords the series an opportunity to discuss British rank and social standing, particularly as it was transferred to the colony. And it also allows a rather cynical interpretation of the false sense of superiority that arose on the part of many of those of British descent.

One of the obvious shortcomings of the Marshall series is the lack of character development. It tends to be superficial, especially of our main characters, Harry Feiffer and Christopher O'Yee. Some facts come out in bits and pieces but not nearly in the manner of Smith or Kaminsky. The readers' attachment to these people, thus, comes not so much from our knowledge of them as complete human beings, but rather from our empathy with them as they solve the crimes in very human ways. One of the strengths of the series is the emphasis on real-life police procedures and the atmosphere of a real-life police station, where a number of cases are going on simultaneously. In some novels, like in *The Far Away Man,* officers are working on a case totally unrelated to the main case; in other novels, like *Thin Air* and *Road Show,* seemingly unrelated cases come together as part of the same crime by the end of the novel.

In *Road Show* real-life procedures parallel the fictional procedures in an eerie manner. Part of the resolution concerns the way in which the wiring on a bomb provides a "signature." We are reminded of the Oklahoma City bombing disaster and the bomb expert from Washington, D.C., who described on CNN how any piece that survives the explosion is important because it provides clues as to who did it and who specifically talked about the way that the wires were bent!

> It was a signature. On every bit of work, no matter how small, there was always the signature of the man who did it . . . Under the glass [Technical Inspector Matthews] had seen the way the bomber had used his watchmaker's pliers to strip the wire to make a connection, then, uniquely, unknowingly, automatically, given it a double twist back on itself and then, turning the pliers in his hand, twisted it back the other way again.
>
> He had done it totally without being aware of it . . . It was a signature.
>
> It was his own. (*Road Show,* pp. 103–104)

Eventually, the technical inspector remembers who saw him do it this way, which helps the investigators identify one of the bombers.

Recent history has furnished the series an intriguing twist with the colony's return to Chinese control in 1997. *Inches, Nightmare Syn-*

drome, and *To the End* incorporate varying degrees of focus on this historic event, with *To the End* taking us right up to July 1997. The blending of real with fictional, once again, affords a powerful way to comment on these changes through an escapist venue. ■■■■■■■■■■■■■■■■■

The Beijing of Christopher West

Capitalism meant selfishness, greed,
disloyalty, immorality.
(Death of a Blue Lantern, P. 42)

IN BEIJING POLICE INSPECTOR Wang Anzhuang, Christopher West has created a post-Tiananmen icon of communist China. Prior to the suppression of the student demonstrators in 1989, he had been a staunch Communist Party member, a war hero even. But Tiananmen had shaken his faith, and the murder investigation he undertakes in *Death of a Blue Lantern* shakes it even more as there are implications of possible party complicity with Chinese gangsters. At the same time, Wang's loyalty to the party is being questioned for his activities during the Tiananmen uprisings.

Chilling reminders of the oppressive nature of the Chinese state add a touch of realism that suggests that, economic reforms notwithstanding, contemporary China is still a totalitarian police state. Post-Tiananmen self-criticism sessions, conducted publicly by the unit committees of the Communist Party, invoke memories of the Great Proletarian Cultural Revolution of the 1960s. Even Inspector Wang is not beyond reproach when it comes to justifying his actions during the Tiananmen uprisings. During the public discussion of his self-criticism in front of the unit committee, his acceptance of a flower from one of the protestors and allowing students to burn campfires to keep warm are called into question. He knows that the easy way out is to accept responsibility and beg forgiveness. All seems well and good until another policeman, noted for informing on colleagues, accuses Wang of trying to interfere with the People's Liberation Army suppression of the rioters. At this point, the stakes become more serious, and Wang realizes that a different tack is necessary.

There is no question that political context provides the major component of our sense of place in this novel. Yet, as with the Japanese series, regionalization is emphasized as a basic element of Chinese society, once again regionalization based on accents: "It had a strong Shandong accent; it belonged to his grandmother, Peng. 'If you learn one thing, it should be that everything changes with time. *Yang* gives way to *Yin*' " (*Death of a Blue Lantern*, p. 37).

Not only do we get a taste of regional variations, but we also get another dose of ancient Chinese wisdom through the philosophy of the Dao. ■■■■■■■■■■■■■■■■■■■■■■■■■■■■■■■■■■■■■■■

Conclusion

These four Oriental police procedural series offer a great variety of techniques and tones for exposing places and underscoring the weaknesses of regional generalization. In just these four, we find communist China juxtaposed against capitalist, colonial China; Japan as interpreted by a Japanese author against the interpretation of a foreign-born author; serious approaches to crime against comedic approaches to the same kinds of crimes.

Two themes appear to unite these series—the tensions between continuity and change and between the similar and the unique. The wisdom of ancient Chinese and Japanese cultures recurs even in the communist-based series, while simultaneously emphasizing the changes that have produced the contemporary societies. Appreciation for the more recent past—World War II, the Cultural Revolution, the American occupation of Japan, to name a few—is also used as a point of reference for understanding the modern era in these three quite different societies (Japan, Hong Kong, and communist China). As well, an appreciation for the great variety in these cultures is often exposed, reminding us that even in China and Japan, two cultures thought to be among the most homogeneous in the world, there is diversity.

> For a second, [Wang] thought about doing what millions of his compatriots had done in the Mao years: breaking down and throwing himself on the mercy of the authorities. Then he thought of what had happened to them. (*Death of a Blue Lantern,* p. 152)

In addition to public rehabilitation, more subtle intrusions of the government appear in the commonplace carrying on of everyday activities: "'We couldn't meet, could we? I mean, I don't trust the phone. People might be listening'" (*Death of a Blue Lantern,* p. 47).

Yet one does not get the impression that Wang has given up on the system just yet. Confused would better describe his predicament. These recent events call certain of his beliefs into question; yet there is no question that a deep-seated faith still holds a powerful grip. He is critical of the younger generation that does not fully appreciate the progress made in communist China since the revolution: "Too many of the capital's young people were like this nowadays—self-absorbed, surly, passionless. Would they really throw away everything the previous two generations had fought, sweated and died for?" (*Death of a Blue Lantern,* p. 11).

Devotion to communism, however, does not preclude an appreciation for centuries of Chinese culture. From time to time, references to the wisdom of the great Chinese philosophers does not seem out of place or necessarily antagonistic to Chinese communism, as long as one doesn't admit to it in self-criticism: "'Words can lie, the eyes image the soul.' . . . Wang smiled—a quote from Mencius: what a splendid, scholarly old fellow!" (*Death of a Blue Lantern,* pp. 39–40).

What we see in Wang Anzhuang is a balance between the continuity of traditional Chinese culture and the changes brought by the introduction of communism. He is a reflection of Chinese communism being as Chinese as it is communist.

> Wang disliked [the students] parroting of this Western word [democracy]. Did they know what it meant? No two of them seemed to have the same definition . . . But another side of him knew that the protestors had a point. There were things wrong with the existing system. Their intentions were good. If only they'd tone down their objections, talk sensibly with the Party—negotiation and consensus, that was the Chinese way. (*Death of a Blue Lantern,* p. 35)

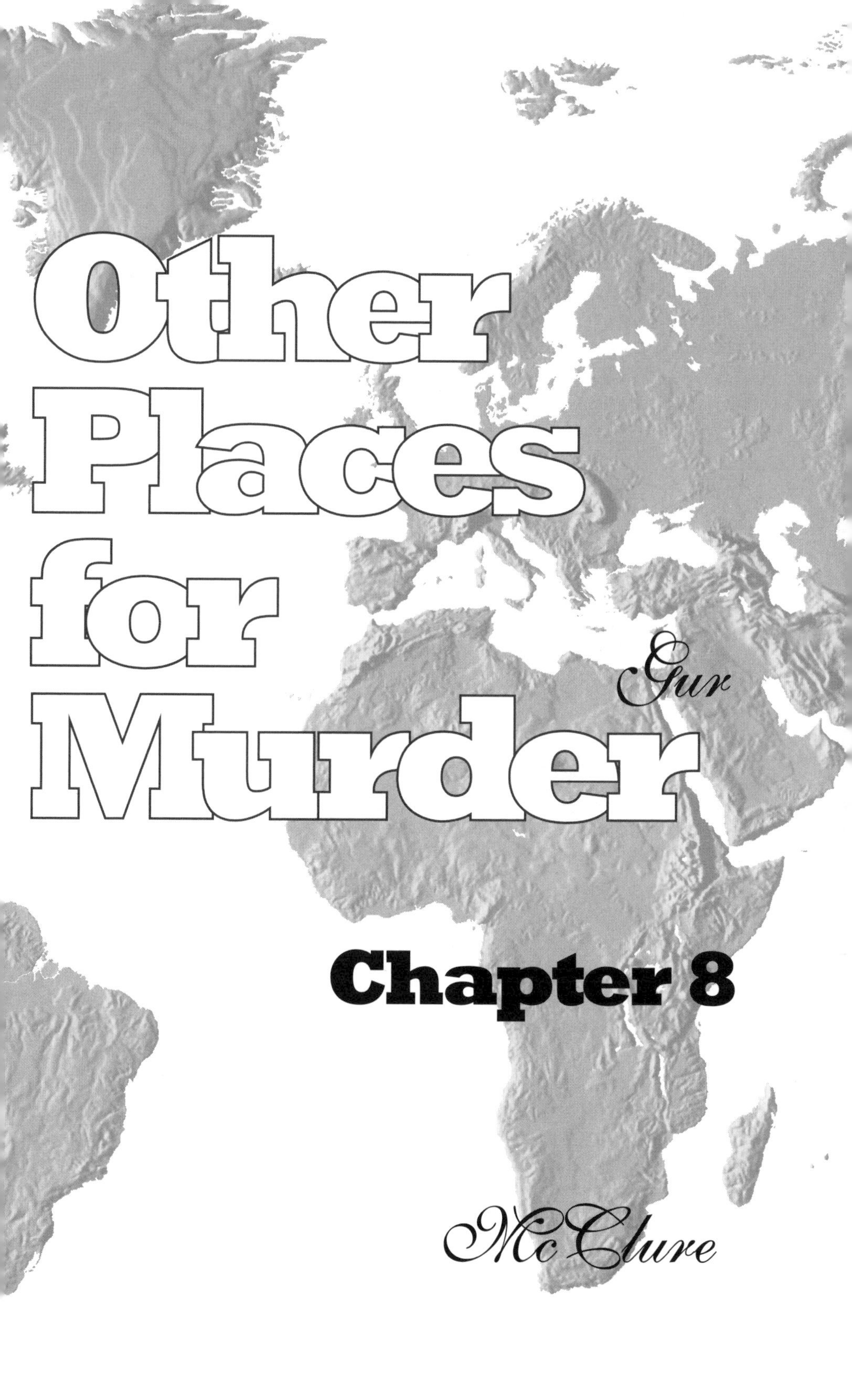
Other Places for Murder
Gur
Chapter 8
McClure

Keating
Upfield

Upon departing the venues of North America and Europe, the selection of places for police procedurals is limited, especially for those available in English. A number of police procedurals have been translated into English from their native languages, but they remain few. Four English-language police procedurals from other places stand out as successful series representative of different approaches to the genre, with much of the differences rooted in the locales where they are set. Three of the series, interestingly enough, are set in the former British Empire. This commonality explains some of the characteristics these series share with the traditional British approach to murder mysteries, not the least of which are the hints of the Great Detective tradition in the protagonists of these series and, of course, the use of English as the official language of communication. At the same time, the uniqueness of places allows the authors to address problem-solving from distinctly non-European perspectives. The other contribution to this collection has quite a different historical and cultural context. Although Israel was under British "protection" between the wars, the legacy of British imperialism is not a major factor, and English is not the language of communication. Israel's role as the world's only Jewish state offers a place for murder that is quite different from its three companions.

The other places selected for this book are Batya Gur's Israel of Michael Ohayon, H. R. F. Keating's Bombay, India, of Inspector Ghote, James McClure's South Africa with Inspectors Kramer and Zondi, and the Australian Outback of Arthur Upfield, who has created Inspector Napoleon Bonaparte.

All four authors carry strong credentials for being able to convey sense of these places. Batya Gur teaches literature in Jerusalem; the series is written in Hebrew and translated into English. Keating's credentials are weakest of the group: for the first ten years he wrote the series without having visited India, which may explain why the series is weakest in narrative description of landscape, more heavily emphasizing cultural and institutional characteristics of place. Our last two authors write about their native homelands—James McClure about his native South Africa and Arthur Upfield about the Australia he immigrated to and where he lived all his adult life. ■■■■■■■■■■■■■■■■■■■■■■■

The Israel of Batya Gur

He knew that he had to immerse himself in the world of these people, knew it was there that he would find the solution.
(Literary Murder, P. 227)

IN THE MICHAEL OHAYON SERIES, Batya Gur has created a uniquely Israeli interpretation of the police procedural, and one that is solidly place-based. Written in Hebrew, the translations into English by Dalya Bilu do not detract from the flow of the prose nor the power of the imagery in these novels. There is none of the stilted style or lost images of many translated works. It is difficult to tell that these novels were not written originally in English.

Israeli society and culture are an essential element in all of the plots, but nowhere is the uniqueness of Israeli society more evident than in *Murder on a Kibbutz,* where it is difficult to tell whether the main focus of the novel is a critique of the kibbutz movement or the resolution of the crime. They are so closely interconnected that a convincing argument could be made for either. What is most intriguing about Gur's approach to the whole idea of the kibbutz is her rather critical appraisal of its value to Israeli society, as conveyed through numerous of the novel's characters. The similarities between the Jewish kibbutz and the Soviet collective farm are not lost on a former kibbutz member during an agricultural celebration: "He could not entirely suppress the feeling that once you took away the blue and white and the flags on the Caterpillar, the whole ceremony seemed archaic and foreign, as if it were taking place on a collective farm in Soviet Russia" (*Murder on a Kibbutz,* p. 4).

Two themes permeate Gur's approach to the kibbutz: first, the disruptive nature of murder on the seeming tranquility of kibbutz life, reminiscent of the rural English approach to literary murder, and the unbelievability that such a deed could have occurred on a kibbutz; and second, the contradiction between the perceived equality and community bonding and the reality of daily kibbutz life. One view to this kind of a society is provided by the series hero, Michael Ohayon, head of the Criminal Investigations Division (CID) in the Jerusalem subdistrict, who is an outsider, having never lived on a kibbutz himself, a point he is reminded of numerous times during the investigation: "'But if you've never lived on a kibbutz,' said Meroz, and Michael knew that he was going to hear this sentence repeated ad nauseam, 'you'll never understand anything'" (*Murder on a Kibbutz,* p. 183).

Yet it is Michael as an outsider who brings the serenity to our attention, a serenity that makes such an impression on an outsider, that he himself questions whether such a diabolical act as murder could ever actually occur here.

> Between one interview and the next, between the thousands of words he had been listening to for the past three days, Michael occasionally caught a glimpse of the stunning tranquillity of the landscape surrounding him. The serenity radiating from the neat paths and lawns, from the playgrounds and the plaza fronting the dining hall, from the cemetery with its separate section for those who had died on military service, seemed to him absurd. It made the whole case seem somehow unreal . . . he wondered whether there had been a murder here at all. (*Murder on a Kibbutz,* pp. 260–261)

The kibbutz movement is brought into question not by an outsider, like Michael, but by insiders, who comment on the conflict and turmoil that actually exists under the facade of tranquility and peace: "'On second thought, it's the very *idea* of the kibbutz that I don't like,' muttered Meroz, as if to himself. 'It's giving too much credit to the human race that it can attain true equality—and among Jews, besides'" (*Murder on a Kibbutz,* p. 185).

As one would expect, there are numerous references to the Holocaust, and many of the characters in these novels have ties to the Holocaust; yet it does not become the main focus of the series, serving mostly as background and history to contemporary Israeli society. In *Literary Murder,* Soviet suppression of Jews becomes a major issue in the plot and key to uncovering the motive for murder. Still, it does not detract from the primary focus of the novel—to solve the murder using good police procedures.

In this series, Batya Gur has truly translated the key elements of the procedural genre into the Israeli case. In Michael Ohayon, we meet a divorced father who struggles with the needs of his separated family while trying to solve murders. To add a realistic twist to the family situation, his son is currently serving in the military, on patrol in "the territories," which in this case refers to Bethlehem. The concern for his son's safety and the omnipresent threat from the *Intifada* are spliced into the story line to remind us of the ongoing reality of daily life for Jews in Israel.

Also in keeping with police procedures, Michael works as part of a team, albeit a team that has its own rivalries and intrigues, especially after Michael is transferred to a new job in *Murder on a Kibbutz.* Gur is careful to create characters that are real for us, some of whom, like Michael, we come to know and like as people, others who are not quite so likable. In all, a group of different kinds of people who work together to solve crimes.

One part of his investigations is the importance of getting the feel for the place as a key to understanding the crime. In *Literary Murder,* Michael makes a strong case for what he calls the "essence of things."

> This business of the "essence of things," which was often mentioned with a smile in all the investigating teams he had worked with, was his personal contribution to an unusual style of detective work. He needed, he felt, to become part of the environment that he was investigating, to sense the subtle nuances of the murdered person's world. (*Literary Murder,* p. 99)

This is more than just a reminder of the importance of getting a feel for places; it's getting a feel for an entire world that is foreign to Michael Ohayon. Even though he is Jewish in the state of Israel, the murders in this series take place in societies not familiar to him—the world of psychoanalysis in *The Saturday Morning Murder,* the world of academe and literature in *Literary Murder,* the kibbutz in *Murder on a Kibbutz,* and the world of classical music in *Murder Duet.* Gur has taken the importance of place a step further within the context of Israeli society, akin to understanding the world of the Navajo within American society in Tony Hillerman novels. The result is further support for the place-based nature of these police procedurals and this series in particular.

The Indian Subcontinent of H. R. F. Keating

'On the one hand,' he said, 'you have an official campaign against bribery. On the other, respected religious writings condone and even encourage it. I do not understand.'
The inspector's face brightened.
'But it is perfectly simple,' [Ghote] said.
'You have explained it yourself. On the one hand, on the other hand. The two things are quite separate.'
(The Perfect Murder, P. 32)

FOR OVER A QUARTER of a century, H. R. F. Keating has provided readers with a uniquely Hindu approach to the world of crime and murder through the exploits of Inspector Ganesh Ghote (pronounced Go-tay) of the crime branch of the Bombay Municipal Police Department. Several characteristics of Keating's approach to sense of place through this series are worthy of attention.

It begins with the main character himself, Ganesh Ghote. Although working alone, with only inconsistent suggestions of being part of a team, Ghote does not follow simply in the footsteps of the Great Detective tradition. In a way, Ghote's personality encompasses both the Great Detective and the dupe in a singular and complex individual. As part of his Hindu approach to crime solving, there is a continuing struggle within Ghote's own mind, often centering on what is right and what is wrong. Unlike many police procedurals where there is no question about right and wrong, for Ghote the question is one of a higher good, even to the point of protecting the murderer. In *The Iciest Sin,* Ghote feels compelled to protect the murderer of a noted blackmailer because of the greater good the murderer serves for Indian society. Ghote is actually a witness to the murder. As might be expected, this betrayal of basic police procedures, not to mention his own basic standards of morality, leads to myriad tests of values, both professional and personal, which raise numerous questions about right and wrong. Many readers of Western police procedurals, used to the basic morality of Western ideas of right and wrong, might have trouble with this kind of an approach to crime.

This approach is reflective of Keating's sense of place in general, whereby our sense of Indian society is not so much derived from descriptions of the urban or rural landscape, which are found only sparingly in the series; rather, they result from dialog and iconography. In this case, Ghote takes a very different view of crime than his Western counterparts would.

It is also in dialog that we are exposed to a sense of India. Keating uses this to capture not only the pattern of English as spoken by many Indians, but the thought processes as well.

> 'Sir, it is a matter of chance only. You see, sir, I was out at Sahar Airport some time ago, obtaining some informations, when I was spotting this girl, one Miss Nicky D'Costa, sir, Goan lady, an air stewardess. She was just only coming out from airport staff security check, and I was straight away noticing her change of manner. From walking in a very very idle way, sir, she was in one moment only starting almost to run.' (*Asking Questions*, p. 16)

This pattern of speech is not just a caricature of Indian speech patterns; rather, it reinforces our impressions of Ghote as a thoughtful, deliberate investigator who, although sacrificing the law to a greater good, does so only after great internal debate and rationalization, properties we are led to believe are indicative of a Hindu approach to problem solving.

Another internal debate identifies Ghote as an icon of the transitory nature of Indian society. *Cheating Death* opens with Ghote in internal chaos over how to deal with his wife concerning who in the family should be making the important decisions. Should he take the traditional Indian approach or is he of a more enlightened era?

> Was he really going to do that? Other husbands did, of course. Other Crime Branch officers even. And cheerfully boasted about

> it . . . Only way to keep the biwi in her place, they said. One good beating every now and again, and no nonsense after about who is making the decisions. Only way to do it. But . . . But was he really going to do that himself? (*Cheating Death,* p. 1)

Just as we are reminded of the changing nature of traditional Indian culture, we are also alerted to the continuing legacies of British colonialism and the caste system, and how difficult it is for some, especially those of privilege and status, to change. In *The Perfect Murder,* visiting Swedish policeman Axel Svensson points out the difficulties of police in former colonies.

> 'This is a great problem,' he said. 'In countries where perhaps the police are associated with the former rulers, when independence comes there will be elements who are not willing to accept that they too must come under the same law as every other citizen.' (*The Perfect Murder,* p. 69)

One aspect of Bombay that does come through is the crowding—the seemingly chaotic overbearing of large masses of people, and the attempts of individuals to survive and persevere. One particular scene finds Ghote meeting with a local crime lord, who is attempting to blackmail Ghote at a train station, where chaos surrounds them. The whole situation serves as a metaphor for the chaos and turmoil within Ghote's soul as he grapples with the moral issues surrounding his impending involvement. His jumping on a train at the last minute to avoid giving in to the temptation symbolizes his decision that he no longer can flaunt the law, even for a higher good. Within a matter of minutes, he makes the transition from potential killer to not allowing himself to be blackmailed. Escape. Peace. Resolution.

> [He] swung himself into, or half into, one of the open doors [of the train] rushing past. The strain on his arm was fearful . . . But then his left foot found a firm grip on the carriage floor, and the strain miraculously eased. (*The Iciest Sin,* p. 160)

One key issue raised as subplot concerns the plight of minority communities and how they work to maintain ethnic pride and cultural values. Historical continuity comes into play once more, this time in the guise of storytellers—another venerable Indian institution. The Parsi storytellers remind Ghote of his own childhood experience with "keepers of tradition and anecdotes of life."

> [Ghote] listened to the tale, remembering boyhood hours spent drinking in long and long stories he had been told . . . It was an account . . . of a pioneer solo feat of aviation . . . Something designed presumably to rouse in its juvenile Parsi hearers a stirring of pride and ambition. (*The Iciest Sin,* p. 78)

A different approach to the police procedural is found in the Ghote series; yet, it is the approach itself that adds to our sense of a very different kind of place. Praised for its handling of the psychological aspects of crime, it has much to offer sense of place. Although much of what we find is exotic, the allure begins with what is familiar—the basic humanity of the main character and the basic ingredients of all police procedurals.

Other important aspects of Indian society emerge when Ghote ventures to the sacred city of Banares in *Doing Wrong.* We are incessantly reminded of the healing powers of the Ganges River, where Hindus go to absolve themselves of sin. Even our murderer believes that he can be cleansed of murder by submerging himself in this most holy of Hindu rivers: "Now that he had shiveringly immersed himself in the dawn-chill waters of the Ganga he was free from that sin . . . Surely, now, all of that had been wiped away. As if it had never happened" (*Doing Wrong,* p. 6).

Not only is Banares the city where Hindus go to cleanse themselves in the Ganges, it is also the holiest of cities to die in. Continual references to funeral pyres and processions throughout the investigation remind us of this. As well, it provides an opportunity to discuss death in India, where on a daily basis many die needlessly of starvation and disease. The murderer tries to use this fact as a rationalization for his own crime. Ghote is not persuaded: "'Oh, yes, sir, I know that the lot of the downtrodden in this country is sometimes very bad, that people are, yes, dying when they should not. But that, wrong as it is, sir, is not murder. Cold-blooded murder'" (*Doing Wrong,* p. 148).

That Ghote and the murderer are having such a conversation, even though Ghote cannot yet prove guilt, is a reflection of another different kind of approach in Keating. In many of the novels, the identity of the murderer is revealed early in the novel. In *The Iciest Sin,* for example, much of the suspense concerns finding out what Ghote is going to do. In *Doing Wrong,* nearly equal time is devoted to the inspector and the murderer, although there is no question that Ghote seeks to bring the murderer to justice, an interesting twist of plot that seems to work well with the overall flavor of Ghote's Hindu personality and approach to investigation. ■

The South Africa of James McClure

They crossed the national road and headed north into the center of Trekkersburg, along streets almost empty of traffic. It was after midnight, long after curfew, and so the only pedestrians were white and few in number.
(The Gooseberry Fool, P. 23)

FOR NEARLY A QUARTER OF A CENTURY, James McClure has been sharing images of his native South Africa, and commentary on its system of apartheid, through the exploits of Detective Inspector Tromp Kramer of the Trekkersburg (Natal) Murder and Robbery Squad and Bantu Detective Sergeant Mickey Zondi.

Several characteristics of this series are particularly powerful in conveying a sense of South Africa. Vivid landscape descriptions give a feel for the place, especially when our heroes venture outside of Trekkersburg. In *The Song Dog,* which describes the original meeting of our two heroes in 1962, the crimes are set in northern Zululand, although a long hike down a mountainside proves that it is possible to get too much of a good thing: "[Kramer had] had more than his fill of picturesque rural scenes typifying Zululand: the mud huts and the aloes, the drought-stunted maize and the potbellied piccanins, the donkeys with rocks tied to their tails" (*The Song Dog,* p. 197).

For the reader, of course, it is not enough—we demand more description to try to imagine this land that is such an enigma to most North Americans and Europeans. McClure does not disappoint, providing vivid images and impressions of scenery and climate often throughout the series.

Yet more enticing for many readers is the human sense of place that so few foreigners understand. McClure begins the explanation by providing brief history lessons throughout, often through the telling of stories. Even characters in the series are interested in trying to understand better the complex history of South Africa.

> "He would come to her because he wanted to learn the story of the Zulu people—she claims her memory goes right back before the Zulu wars with the English, and that her father was at the kraal of Shaka when the Voortrekker leaders were killed there after being warmly welcomed." (*The Song Dog,* p. 192)

For McClure, the use of historic perspective includes another aspect as well. Actual historical events are alluded to, events many of the readers can relate to, in order to keep the actions of various characters in perspective. In *The Song Dog,* which recounts the case that originally brought Kramer and Zondi together, the arrest of Nelson Mandela at one point occupies the time of the detectives' chief back in Trekkersburg.

> "Captain Bronkhurst was busy with a very big investigation, assisting the Security Branch to find a certain Bantu male, Nelson Mandela."
>
> "Who?" asked Kramer.
>
> "Oh, some Xhosa," said Zondi with what seemed like a very Zulu gesture of dismissal for someone belonging to a lesser tribe. (*The Song Dog,* p. 163)

The result of this complex history, of course, is a society consisting of at least four basic ethnic groups—English descendants, Afrikaners (descendants of the Dutch), Zulu aborigines, of which the Bantu is one of hundreds of different tribes, and coloreds (persons of mixed race). Westerners tend to associate apartheid with segregation solely between whites and blacks; yet the mixed-race coloreds often are perceived by white South Africans just as derogatorily as blacks: "Many whites believed coloreds were a mixture of the worst characteristic of all the races whose blood ran in their veins" (*Snake,* p. 116).

Ethnic antagonisms and the racial system of apartheid that arose is a major focus of these novels. McClure admits that influencing the opinions of older South Africans, who read murder mysteries in great numbers, was one of the main reasons he chose the police procedural as the particular genre for carrying his message. If much of the sense of place found in escapist literature is insidious and not noticeable, so is McClure's message about apartheid. He delivers his message quite successfully, never seeming to preach or proselytize, although his particular bias is unmistakable, provided through the perspectives of the characters in the series.

There is no question that through the eyes of the Afrikaner Kramer, there are serious concerns about the morality of apartheid. Yet it is done often only in passing, as an aside to the basic narrative.

> In fact the fast section lasted only as long as the length of [the Bantu township of] Peacehaven. It took the vulnerable white motorist through as quickly as possible, reducing the shacks and shanties to a colourful blur, and provided an excellent surface for the deployment of military vehicles in the event of a civil disturbance. (*The Steam Pig*, p. 73)

But the strength of this particular aspect of the series derives from the interplay between Kramer and Zondi, Afrikaner and Bantu, another case in which sense of place results from the combination of iconography and dialog. It is through Zondi that basic characteristics of Zulu culture are revealed.

> "Heaven, boss. That is what 'Zulu' means: the people of Heaven."
>
> "Christ, they never taught us *that* in bloody Sunday school!" said Kramer. (*The Song Dog*, p. 175)

In this rather simple use of dialog, not only is a basic definition of Zulu given, but a basic lack of understanding by the dominant white culture is revealed as well, from a perspective that is explicitly critical of the dominant culture's oversight of the indigenous peoples. Also present is the nature of the relationship between blacks and whites, even professionally. All white officers are referred to as "boss" by their black counterparts. And there are separate departments for murder and robbery and Bantu murder and robbery.

Whereas Kramer presents a sympathetic white perspective to black culture, Detective Sergeant Zondi presents the black perspective directly, often with a bit of humor—some might say black humor.

> [I]t was as though most white murderers felt they had a tradition to maintain, certain standards to uphold . . . Or was it because [whites] tended, in the main, to be less passionate, less impul-

> sive, and far more cold-blooded in their killing.(*The Song Dog,* p. 181)

Or possibly it's because of basic differences in the approach to life taken by the black and white cultures.

> "Tell me, when the Almighty made [blacks], did he give them souls, hey?"
>
> "The boss means the same as the white man?"
>
> "Uh-huh, of course."
>
> "*Hau,* God would never do such a terrible thing, Lieutenant." (*The Song Dog,* p. 252)

Little humor, however, is found in the policies of resettlement of entire villages to "black homelands": "it was an eviction. An ordinary Black Spot eviction, one of hundreds, an everyday event" (*The Gooseberry Fool,* pp. 39–40).

The effectiveness of McClure's commentary on this aspect of South African apartheid derives from its role in the plot. The investigation of the murder is hindered by this resettlement, whereby a houseboy's disappearance can be interpreted as an admission of guilt or as necessary to help his family during this difficult time. It is presented as a matter of fact, but the pain and anguish caused by resettlement is portrayed through the individual hardships Zondi comes in contact with in the "homeland."

Race antagonisms are not limited to black versus white. Hatred between Afrikaner and English is also a part of South African society, as depicted in the case of a white Afrikaner police officer in *The Song Dog,* when he tries to prevent the marriage of a woman he loves and covets to an Englishman with a touch of Jewish blood. The antagonism between Boer and English provides the rationale for the murder and an important clue once it finally comes to light.

> [He] acted like a—well, the way he hates everyone who isn't Afrikaner *and* Nationalist Party is bad enough, but English-speaking stuck-ups who went to private school . . . He started talking about racial purity, the need to honor the Boer Nation in spirit and deed. (*The Song Dog,* p. 280)

In addition to his major concern for expressing the injustice of the apartheid system, McClure also depends heavily on the use of maps in the solving of crimes. Maps on the walls are a must to get a sense of spatial relationships. But also of great importance are the use of mental maps.

> It was a good thing Zondi had spent so much time staring at the ground during the flight from Mabata to the sorcerer's mountain. Without the mental map he had made, the confusion of footpaths . . . would have led to a great deal of wasted time and energy. (*The Song Dog,* p. 196)

In this particular series, there is no question that McClure has a not-so-hidden agenda to influence attitudes toward the South African policy of apartheid. Yet this does not diminish his ability to provide vivid and insightful images of South Africa. If anything, it enhances the sense of place, if for no other reason than the need to gain credibility in the eyes of those he hopes to affect—people who have first-hand knowledge of South Africa and who will be his severest critics should he get it wrong. As a result, we all benefit from his need to provide an authentic sense of South Africa. ■■■■■■■■■■■■■■■■■■■■■■■■■■■■■■■■■■■

The Australian Outback of Arthur Upfield

"Without doubt you are familiar with the geography of this north-west corner of Australia."
(The Will of the Tribe, P. 28)

ARTHUR UPFIELD IS AN EARLIER, Australian version of Tony Hillerman, with whom he is compared often. Possibly, it would be more accurate to say that Tony Hillerman is a latter-day Arthur Upfield. Clearly, Upfield was a pioneer in the field. In twenty-eight novels, dating from 1929 to 1963, the series follows the career of Detective Inspector Napoleon "Bony" Bonaparte of the Queensland Police, Criminal Investigation Branch. The settings for the series include the wide-open spaces of western Australia's Outback, with a number of ventures to other regions of Australia. Whereas McClure explores racial and ethnic antagonisms through the perspectives of two policemen, one aboriginal and one white, Upfield unites these tensions in one person. In the person of Inspector "Bony," we find investigative traits that combine two literary traditions—the Great Detective and the police procedural, thus borrowing from both Upfield's British roots and the more realistic practices of everyday police work. It may be stretching the definition a bit to refer to this series as police procedural, a genre that would not be "officially" introduced into popular literature until the mid-1940s.

Great Detective, police procedural, or some combination thereof, the popularity of the series to a great extent depends on the main character, who comes across as a believable and compelling personality. Over the course of the series, we come to know "Bony" quite well. His half-caste appearance is striking.

> [S]he studied the dark brown face on which aboriginal race moulding was absent. The face was neither round nor long. The nose was straight, the mouth flexible. The brows were not unlike a verandah to shadow the unusual blue eyes, and although the black hair was now graying at the temples, it was virile and well kept. (*The Will of the Tribe,* p. 27)

To enhance the iconographic power of the series, the most important part of Bony's character is that he is half-caste, which has not only created this unusual personality, but is also the reason he is so successful as a policeman: "the inherited influences of the two races warred for the soul of Napoleon Bonaparte" (*Man of Two Tribes,* p. 57).

The result of this internal war is a "man who could think like an aborigine and reason like a white man" (*The Will of the Tribe,* p. 38). In addition to providing Bony with the skills necessary to solve especially difficult cases, it also gives him an endearing personality.

> [Bony] melted Major Reeves's reserve, which his duality of race had created, with his cultured voice, his winning smile, and his vast store of knowledge . . . He charmed [the Major] as he charmed everyone after five minutes of conversation. (*Murder Down Under,* p. 10)

One of Bony's primary assets is his ability to use to his advantage whichever side of his "duality of race" is necessary for solving the case. Often his aboriginal side is of great benefit, especially in the Outback where he is called upon more than a few times to track, or to trek, or just to survive. On the other hand, his abilities are often underestimated by Europeans who assume that he is nothing more than a backwoods aborigine. Bony uses this tendency to underestimate his skills to advantage as well. Bony himself, however, can never be accused of modesty about his abilities as a criminal investigator, although he treats his success as a matter of fact, rather than as a source of some sense of superiority: "'I am at the top of my chosen profession . . . With never a failure to [my] record . . . I have found my road in my own way, at my own pace, and no one tells me to do this or that'" (*Man of Two Tribes,* pp. 20–21).

Like Hillerman's novels, one of the strengths of the series is landscape description, often coupled with local folklore and mythology, and often sheltering clues important to the resolution of the crime. Western Australia provides ample opportunities for all three. Long distances between places are taken for granted, as are the challenges associated with such distances in the harsh environments of this part of the continent: "'in this part of Australia three miles is reckoned as being just outside the back door'"; "Here is a land where distance is measured by the hour, where only the initiated can hope to move from one point to another, and where only the bush masters can find water" (*Man of Two Tribes,* pp. 26 and 43).

The terrain we encounter is the vastness of the Australian Outback—desolate, foreboding, unforgiving, and intimidating, yet with a beauty of its own, with an appeal to a certain kind of person, and to the aborigines. The Nullarbor Plain serves as the setting for *Man of Two Tribes.* Surviving in the plain becomes a primary necessity for the resolution of the mystery. Its vastness is introduced right off the bat.

> On a moonless night there is nothing to be seen of the Nullarbor Plain, or of the railway which crosses it for three hundred and thirty miles without an angle Euclid could detect, nothing of all those square miles of table-flat, treeless land beneath which the aborigines believe Gamba still lives and emerges at night . . . Now were hidden all the caves, the caverns and the blow-holes, and the miles on miles of foot-high salt bush. (*Man of Two Tribes,* pp. 1–2)

The connectivity of landscape description, aboriginal folklore, and clues for solving the mystery are supplied from the very outset of the narrative. Immediately, we are alerted to the fact that the topography of this uninviting locale and an understanding of its place in aboriginal mythology are going to play pivotal roles in the mystery as it unfolds. Upfield is a master at getting this out in the open for the reader early on in the process of weaving his tale of murder. He conveys a feeling of fear and respect that he developed over the course of his adult life spent in this part of the world. This fear and respect are transferred to his detective as well, along with the ability of an aboriginal to effectively deal with and survive this kind of environment. In the series, we often see a transformation as Bony changes from the suave, urbane personality of his European half to the cunning, desert-smart personality of his indigenous half. The ability to make such a transformation is the key to his success as a criminal investigator. Unlike many, Bony is proud of his half-caste origins, which he recognizes as the strength of his personality. This is offered in stark contrast to the murderer, Rex, in *No Footprints in the Bush,* whose motive for killing is related to his hatred for his father, because Rex is half-caste. His whole life has been spent trying to gain status and respect in a white-dominated society.

Bony's half-caste status provides Upfield the opportunity to address a number of related issues, including a basic understanding of indigenous culture and history and the nature of the relationship between blacks and whites in Australia.

Bony's knowledge of indigenous culture is essential to solving many of these cases. In *No Footprints in the Bush,* understanding the importance of aboriginal mythology is necessary to solving the case, reminiscent of Hillerman's story lines. It also helps to provide sanctuary when being chased by a band of local aborigines: "the Illprinka men had halted, and stood staring at them. The great stone stood guard over the tribe's sacred treasure house, and even within its shadow there must be no violence. In the shadow of the stone was sanctuary" (*No Footprints in the Bush,* p. 172).

Bony also takes opportunities from time to time to remind readers that although much of aboriginal culture appears to be backward and superstition-laden, in many ways it is quite sophisticated. In *The Will of the Tribe,* Bony knows that local aborigines know how a body got into the meteor crater, even though they initially deny any knowledge.

> "[The dead man] couldn't cross the desert without the wild aborigines knowing all about him . . . the aborigines still removed from long and close association with the whites also have a broadcasting system developed through past centuries, and I

> think you will agree with me, too, that espionage organizations set up by outside governments are amateurish by comparison with the methods employed by these Australian aborigines." (*The Will of the Tribe,* pp. 28–29)

For the most part, white society has not bought into the idea of appreciating indigenous cultures, traditions, or methods, thus creating a basic, if not all too familiar, environment of white domination and racial tension.

> "The white man says . . . I own the country, and these black people must be raised up and integrated with my civilisation. He actually believes that his way of life is better than ours: he can't understand that we don't want to be dragged from the Garden he once lived in, we don't want to be dragged down to his level." (*The Will of the Tribe,* p. 80)

One of the truly endearing, if not frustrating, aspects of European Australian culture is the rich terminology that has evolved in this realm of the former British Empire. The free use of terms, without definition or explanation, such as "walkabout," "tuckerbox," "pannikin," only adds to our sense of the diversity of the English language, recognizing its peculiarly Australian variant, even if we don't always understand precisely what the words mean.

Upfield also comes across as a good geographer, or at least an appreciator of the need for geographical knowledge. In some cases, to help the reader picture the spatial relationships he or she must understand in order to solve the mystery, Upfield provides a map, as he does in *Murder Down Under,* where a map of the town of Burracoppin assists us from the beginning as we follow the unraveling of the mystery.

Although racial issues serve as the primary set of social issues for this series, numerous other social issues arise as well, although not as consistently. As an icon with roots in both cultures, Bony provides a perspective born of both traditions that sometimes grapples with basic issues of right and wrong, and it is not always a simple case of being right or wrong, especially when faced with the human side of crime. One of the innovations of the police procedural genre was to make the criminals human beings as well, with reasons and motivations for their crimes, other than simply being evil people. Upfield has incorporated this aspect of the genre into Bony's personality.

> Here were six murderers, and here was he who loathed murderers and hunted them relentlessly. Right now could he hate [them]? . . . This hunt . . . which had led to the discovery of a community of murderers, had become personal.
>
> There was not among them a human tiger beyond reformation.
>
> He must beware of such reflections lest they should influence his approach to future man hunting. (*Man of Two Tribes*, p. 135)

The popularity of the series, actually written in the late 1920s through the 1960s, attests to the timelessness and appeal of the issues raised by Arthur Upfield, who in many ways was ahead of his time. Seen together with the other series discussed in this chapter, it shares with them several characteristics. All four of these series have been, some still are, prolific and popular, presenting excellent examples of escapist literature. For the three based in former colonies of the British Empire, racial and ethnic antagonisms are the main social issues raised. For our other series, the evolution of an Israeli society in a Jewish state serves as the focal issue. Most importantly, all are strongly place-based, providing vivid images of murder in other places. ■■■■■■■■■■■■■■■■■■■

Conclusion

These series expand the field to include quite diverse cultural and geographic contexts. Emphases change. Racial tension is a more visible subplot; historical background takes on added meaning for setting the plot. Yet these novels still display commonalities: all these series have compelling protagonists, they employ police procedures to resolve murders, and place is an essential plot element. This chapter reveals the ability to expand the genre geographically. We next look to extend the genre historically.

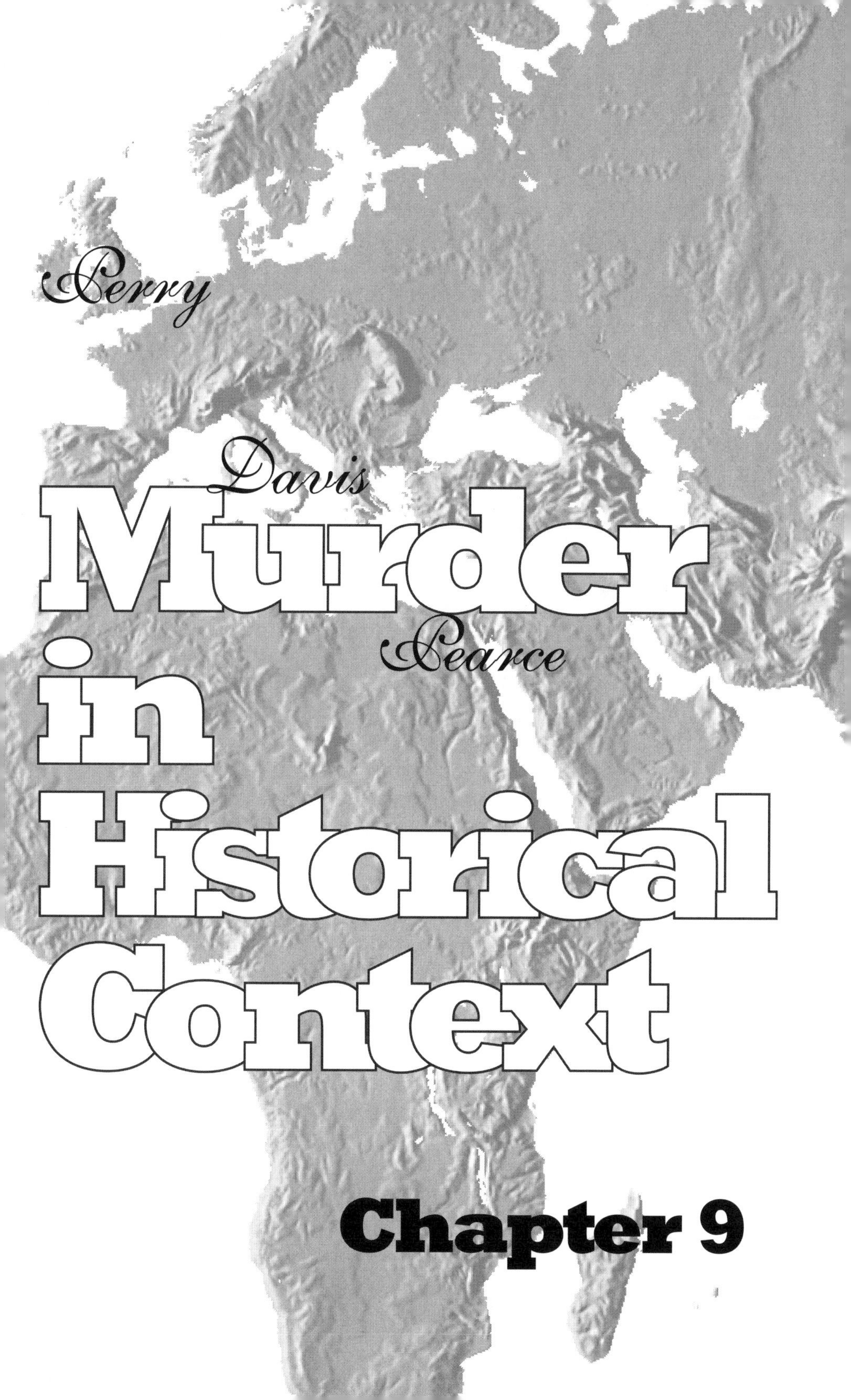
Perry
Davis
Murder
Pearce
in
Historical
Context
Chapter 9

Van Gulik

An intriguing twist to the police procedural is provided by those authors who transport us not only in place, but in time as well, writing murder mysteries set in the distant past, from ancient Rome, to seventh-century China, to Victorian England, to turn-of-the-century Cairo. This poses an even greater challenge to the author to get the procedures right for the given period in history. The four authors described in this chapter are representative of such efforts, and we take them in turn, beginning with the most distant past and working toward the start of the twentieth century. We begin with the Imperial Rome of Lindsey Davis and her main character, Marcus Didius Falco, described as an "informer" who works periodically for the emperor. The series takes place in the first century A.D. From the Roman Empire we move eastward geographically and six centuries in time to explore seventh-century China through the exploits of Robert Van Gulik's Judge Dee. Anne Perry's Victorian England is representative of a number of mystery series set in this time and place. She has created Inspector William Monk, who will later become Private Investigator Monk. We conclude our exercise in historical sense of place in turn-of-the-century Cairo with Michael Pearce's Mamur Zapt.

For the better part of this chapter, it is necessary to suspend rigor in applying our definition of police procedurals. Only Anne Perry's William Monk is actually a policeman and in only the first two novels at that. On the other hand, each investigator in his official capacity, employed by the government, solves murders using investigation techniques that are procedural in orientation, giving the impression that they could have been real-life procedures for the time and place involved. We have stretched our definition so that we may briefly examine examples of creative and innovative ways for conveying sense of places that are exotic both geographically and historically. ■■■■■■■■■■■■■■■■■

The Roman Empire of Lindsey Davis

Changing Emperors had been fashionable for the past two years. After Nero's numbing concerts lost their appeal.
(Silver Pigs, P. 66)

SETTING HER NOVELS in the Roman Empire during the latter part of the first century A.D., Lindsey Davis has attempted to transpose the hard-boiled detective genre to this historical setting. Loaded with a humorous, sometimes farcical approach to her main character, the story line is appealing to modern audiences who can relate to these kinds of character traits, even if the settings and cultures are so foreign.

The decision to include this series rests on two overriding considerations. One, there were no police in Imperial Rome, and thus it is impossible to create a police procedural for this period. Referred to as an "informer" in first-century Rome, the hard-boiled detective types, working for the emperor, are the closest we can hope for. The more important reason, however, is the rich description of place provided by the series. In this respect, the series is unique in offering a sense of place for this era through escapist literature.

The development of the main character, Marcus Didius Falco, is reminiscent of the television character Jim Rockford as portrayed by James Garner several years ago. Like Rockford, the Falco character is a modernized, more human version of the traditional hard-boiled private eye. The most endearing quality of the character is that he doesn't take himself seriously all the time and is prone to mistakes and mishaps. In a self-deprecating, amusing, and endearing way, Falco has no illusions about his profession. When asked what being an "informer" was about, he replies:

> "Information mostly. Finding evidence for barristers . . . or just listening for gossip, more often than not. Helping election candidates slander their opponents. Helping husbands find reasons to divorce wives they've grown tired of. Helping wives avoid paying blackmail to lovers they've discarded." (*Last Act in Palmyra,* p. 7)

Yet, from time to time, Marcus takes jobs for the emperor himself, Vespasian Caesar, who has come to rule after the civil wars of the first century A.D., and most often these assignments involve murder. Falco takes these assignments even though he is ardently opposed to the imperial system: "whenever I had more creditors than usual, or when I forgot how much I loathed the work, I agreed to imperial employment. Though I despised myself for becoming a tool of the state, I had earned some cash" (*Last Act in Palmyra*, p. 9).

It is through the Falco character, who is an ardent republican, that Davis is able to slip in political commentary about the imperial system.

> "Why did your friend call you a tricky character?"
>
> "I'm a republican."
>
> "*Why* are you a republican?"
>
> "Because every free man should have a voice in the government . . . Because the senate should not hand control of the Empire for life to one mortal, who may turn out insane or corrupt or immoral—and probably will." (*Silver Pigs*, p. 17)

Our images of Rome in the eighth decade A.D. begin at the Forum, where Davis' descriptions reveal a scene familiar to many who have seen images of the ancient ruins of Rome. Yet it is through fiction that she brings the place to life, reminding us that these places once witnessed a bustling and lively society. These places are transformed from soon-to-be museums into real places, built originally for real people and a real culture. Imperial Rome comes to life for us as we are exposed to the sights and sounds and smells of daily life.

> It was the usual scene in the Forum. We had the Record Office and Capitol Hill hard above us on the left; to the right the Courts, and the Temple of Castor further down the Sacred Way. Opposite, beyond the white marble rostrum, stood the Senate House. All the porticos were crammed with butchers and bankers, all the open spaces filled with sweaty crowds, mainly men. The piazza rang with the curses of strings of slaves crisscrossing like a badly organized military display. The air simmered with the reek of garlic and hair pomade. (*Silver Pigs*, p. 4)

One way in which Davis increases our sense of realism is the continuous use of references to actual historical events. Passages from Roman histories are often used to introduce parts of the book. In the first mystery of the series, *Silver Pigs*, Davis leads off the introduction with a brief history of recent unrest in Rome, which provides the backdrop for the story to follow. Falco uses a trip to Britain to think about who would be trying to depose the Emperor Vespasian and, in so doing, gives us some insight into the struggle for control of the empire, describing the free-for-all that followed Nero's demise.

> First Galba, a doddering old autocrat from Spain. Next, Otho, who had been Nero's ponce . . . After him, Vitellius, a bullying glutton.
>
> All that in twelve months. It was getting to seem that anyone with half an education and a winning smile could persuade the Empire that purple was just his colour. Then . . . up cropped this canny old general Vespasian. (*Silver Pigs,* pp. 66–67)

Characteristically, the historical perspective is given with a bite of sarcasm, a trait that dominates Falco's approach to the recent political situation and the events leading up to it. This sarcastic approach carries over to his descriptions of everyday life in first-century Rome, where attention to detail provides the legitimacy required to pull off the sense of authenticity so necessary to make the series work. It is in the mundane, where Falco lives and works, that we are confronted with some idea of what it might have been like living in Rome as an average citizen. Exposure to how the wealthy and powerful lived comes from Falco's dealings with the emperor's court and the senators in his role as informer. We are reminded that in earlier times, even the most basic functions could be difficult, especially if you lived on the sixth floor. We are also reminded that these same basic functions can serve as acts of political defiance as well.

> "I need to go to the lavatory again."
>
> "There are several alternatives. You can pop downstairs and try to persuade Lenia to open up the laundry after hours. Or you can run along the street to the big public convenience—but don't forget to take your [money] to get in."
>
> "I suppose," snapped Sosia haughtily, "you and your men friends pee off the balcony?"
>
> I looked shocked. I was, mildly. "Don't you know there are laws against that?"
>
> I peed off the balcony just to prove my independence. (*Silver Pigs,* p. 18)

On a number of occasions, Marcus Didius is called upon to leave Rome for the outlying regions of the empire. In *The Iron Hand of Mars,* he ventures north to Gaul and Germany. In those novels where Falco does travel, Lindsey provides a map of the empire at the beginning of the book, so that the reader can trace Falco's journey. And as if that isn't sufficient to make a case for the importance of maps, we are reminded that maps were critical tools for travel even two thousand years ago: "I settled down . . . to study my maps . . . the Palace had equipped me with a first-rate set of military itineraries for all the major highways—the full legacy of seventy years of Roman activity in central Europe" (*The Iron Hand of Mars,* p. 50).

As he ventures out into these regions, we are reminded of just how foreboding such trips must have been, especially traveling into those areas that were still blank spaces on the map. We also get a quick lesson on cultural differences as we move away from the heart of the Roman Empire into the extremities. Approaching the Rhenus River:

> The terrain looked decidedly foreign now. Instead of huge Italian villas with absentee landlords and hundreds of slaves, we were riding past modest tenant farms. Pigs instead of sheep. Fewer olive groves and thinner vineyards with every milestone . . . Everywhere was colder, wetter and darker than when we had left home. (*The Iron Hand of Mars,* p. 57)

In *Last Act in Palmyra,* Falco ventures southeastward to the territory of Nabataea on the Arabian peninsula, just to the east of Judea and Syria. In addition to giving Davis another venue for commenting on the spread of Roman imperialism, it also gives occasion to provide landscape descriptions that are in stark contrast to Rome and to earlier journeys to northern and western Europe. Added to earlier descriptions of continental, maritime, and Mediterranean environments are desert realms that elicit positive impressions both of the environments and the manner in which the inhabitants have adapted to them. We are caught up in Falco's surprise and delight at the newfound territories of what will later be known as the Middle East.

> We noticed that although this was a desert, there were gardens everywhere. The Nabataeans possessed spring water, and made the most of conserving rainfall . . . All the same it *was* a desert; when it did rain on our journey, a shower had covered our clothing with fine reddish dust. (*Last Act in Palmyra,* p. 32)

The vast expanses of the Roman Empire have now been exposed to twentieth-century readers, exposed and interpreted through the perspective of a first-century Roman, for whom most of this is new as well, thus providing a rather unique introduction to many of these places for imperial murder. ■■■■■■■■■■■■■■■■■■■■■■■■■■■■■■■■

The Seventh-Century China of Robert Van Gulik

Van Gulik preserved extraordinarily well the way of life of imperial China . . . The smallest items— ink stones, nails in a Tartar shoe, the gongs of Taoist monks, door knobs—are brought into the stories at strategic points in the plot to give Van Gulik the opportunity to enlighten the Western reader about these strange objects and their functions.

(DONALD LACH, INTRODUCTION TO *The Chinese Bell Murders,* P. 11)

POSSIBLY THE MOST CHALLENGING of all mystery series from the perspective of the author is Van Gulik's Judge Dee series. Van Gulik was a Dutch diplomat and scholar who discovered Chinese detective fiction from as early as the seventeenth century—so much for Edgar Allan Poe creating the first fictional detective. Donald Lach's Introduction, which is repeated in all the mysteries, provides insight into the incredible true-life story of Van Gulik himself. Educated in Asian languages and studies, Van Gulik was a member of the Dutch Foreign Service, serving in various Asian posts throughout his career. He was a scholar of Chinese literati. The seventeen novels he wrote between 1949 and 1967 provide "one of the best available means for recovering a bit of the everyday life of [China's] past" (Donald Lach, Introduction to *The Chinese Bell Murders,* p. 1). Some of the earlier plots were taken from Chinese stories; for most, Van Gulik supplied themes and plots, pulling from his extensive knowledge of Chinese culture and society for setting and place. His first book, *Celebrated Cases of Judge*

Dee, was a straightforward translation of an eighteenth-century Chinese text. Because of the very different structure of these early Chinese stories from traditional Western whodunits, Van Gulik substituted his own story lines, based on Chinese sources about a real-life seventh-century Chinese magistrate—a position that included duties far beyond those we normally associate with a judge. One of those duties was to carry out criminal investigations, which loosely qualifies the series as a police procedural. Although license is used in borrowing cultural traits from various centuries and dynasties, the strength of the series lies in its depiction of daily life in Imperial China. "He also enlivened the novels with his own imaginary maps and with his drawings of Chinese scenes based on sixteenth-century pictorial block prints" (Donald Lach, Introduction to *The Chinese Bell Murders,* p. 11).

Although character development is not especially detailed in the series, we are drawn to the characters because they are made to seem human, with human weaknesses and foibles. On the other hand, there is a strong emphasis on teamwork. Because the duties of a magistrate like Judge Dee are so wide ranging, it is essential to have a team of investigators, referred to as his "detective force." Four members of his detective force reappear often enough in the series to be designated as regulars. Two of them are killed, Liang Hoong in *The Chinese Nail Murders* and Chiao Tai in *The Monkey and the Tiger,* and two of them are married off, Ma Joong in *The Willow Pattern* and Tao Gan in *Murder in Canton,* during the course of the Judge Dee adventures. Judge Dee himself takes a wife, his "Third Lady," during the series.

Herein lies one of the lures of the series—exposing us to ancient Chinese culture and society, especially to those aspects of it that are so different from the modern and/or Western experience. That ancient China had to deal with all sorts of criminal activities and that crime was a part of daily urban life suggests that certain negative aspects of cities haven't changed all that much over the past thirteen hundred years: "'Those despicable ruffians roam all through the town like hungry dogs. If in a dark alley they happen to meet a defenceless old man they knock him down and rob him of the few strings of copper cash he carries'" (*The Chinese Bell Murders,* p. 90).

Added to these narrative descriptions are the graphics provided by Van Gulik himself, who drew them "in the Chinese style." These include both scenes from the stories and maps at the beginning of the stories so that we might follow the intricacies of the investigation, several parts of which might be occurring at the same time.

One aspect of Chinese society that Westerners may find surprising was the influx and influence of Arabs in southern China. In *Murder in Canton,* one of the few novels set in a real city, the presence of a large Arab enclave is background for Judge Dee's inquiries into the disappearance of a court official. We are reminded of the spread of Islam in the seventh century into southeast and east Asia, where it was accepted less than enthusiastically by local inhabitants. What we tend to forget is that Arab traders were responsible in great measure for the spread of Islam, and the reason it was so successful in finding root in places like China was because Arabs merchants were good at their craft, which proved to be of benefit to the local economy as well.

The presence of Arabs also gives Van Gulik an opportunity to comment upon the sense of superiority that the Chinese had in relation to other cultures.

> 'It's Arab writing.'
>
> 'How many letters do you have?' Chiao Tai asked Mansur.
>
> 'Twenty-eight,' he replied curtly.
>
> 'Holy heaven,' Chiao Tai exclaimed. 'Is that all? We have more than twenty thousand, you know!'
>
> 'How in hell can they express their thoughts in only twenty-eight letters?' (*Murder in Canton,* pp. 47–48)

Of particular interest to the series is the nature of the Chinese legal system and the role of a chief magistrate within the system. As we discover through the exploits of Judge Dee, the magistrate is responsible for gathering the evidence, often by using clandestine methods; conducting the interrogation, in which torture may be used to extract a confession which is necessary according to Chinese law; and deciding upon the punishment for the crime. Some of this may seem a bit barbaric to Westerners; however, the use of torture is contemplated only after an extensive investigation has made obvious who the criminal is.

> 'It is now perfectly clear to me,' [Sergeant Hoong] said, 'that Your Honour's reasoning is based on solid facts. When the criminal is caught, there is ample evidence to confront him with and to make him confess, if necessary by applying torture.' (*The Chinese Bell Murders,* p. 91)

In essence, the judge is already convinced of the person's guilt. The interrogation and admission of guilt is, for all intents and purposes, a formality but one that is required of the process: "'This person confesses his guilt!' the judge repeated, giving the formula required by etiquette" (*The Chinese Lake Murders*, p. 200).

In part the appeal of the Judge Dee series is the fact that, very loosely, the stories are based on a real-life character. In his research, Van Gulik discovered that the real-life Dee Jen-djieh was revered as an "ideal district magistrate" under the reign of the Empress Wu, during the Tang dynasty. Many of the traits of dynastic China used in the novels, however, are drawn from later periods, especially from the Ming dynasty; thus, it is more correct to characterize the cultural-historical context as "a generalized medieval China." Because this generalization is based on a real person and real places only adds to the importance of place in the Judge Dee series. ■■■■■■■■■■■■■■■■■■■■■■■■■■■■■■■■■■■

The Victorian England of Anne Perry

"You can hardly equate the death of
a marquis's son with that of some thief
or indigent in the street!" she snapped back.
"Nobody has more than one life to lose, ma'am;
and all are equal before the law, or they should be."
(The Face of a Stranger, P. 78)

OF THE NUMEROUS Victorian-era mystery series, Anne Perry's Inspector William Monk series provides the best example of the police procedural at a time and place so closely associated with the mystery genre. Although Monk leaves the police force after the second novel to pursue a career as a "private inquiry agent," the first two novels offer sufficient examples, as do those written after his departure from the force, of the effective use of sense of place.

Perry is well known for her character development, and in William Monk, she has created a compelling protagonist. We are introduced to Monk as he regains consciousness after a serious accident from which he has lost all memory—he has absolutely no recollection. We learn with him as he slowly discovers who he is. This serves as an ingenious tool for character development, especially because Monk himself experiences the unveiling of his character as an outsider, often disheartened, or at least disappointed by the kind of person he was apparently prior to the accident.

> He admired the man he saw reflected in the records, admired his skill and his brain, his energy and tenacity, even his courage; but he could not like him. There was no warmth, no vulnerability, nothing of human hopes or fears, none of the idiosyncrasies that betray the dreams of the heart. (*The Face of a Stranger,* p. 116)

Both his knowledge of who he is and his investigative skills return over the course of these first two novels as he solves murders with the help of another compelling character—the brash and liberal-minded nurse, Hester Latterly, who had served with Florence Nightingale in the Crimean War. Thus, two compelling characters form the foundation for the series, and their love-hate relationship will last even after Monk leaves the police force: "Hester Latterly, [was Monk's] sometimes friend, sometimes antagonist, and frequent assistant, whether he wished it or not" (*A Sudden, Fearful Death,* p. 2).

A third character who reappears throughout the series is the young policeman, John Evans, who becomes Monk's partner after the accident and is one of the few people that Monk genuinely likes. Evans will remain on the force after Monk's departure but shows up in subsequent novels as the officer assigned to the cases Monk and Latterly get involved with.

Not only are all these characters appealing in their own rights, they are also social icons of the late-Victorian era. Whereas Monk represents more traditional, male-dominated English society of the period, Latterly represents the newly emerging female voice. Evans, on the other hand, represents a younger version of Monk with all of those character traits Monk discovers were lacking in himself prior to the accident. Evans heralds the changing nature of masculinity in Victorian England.

If Inspector Monk spends much of his time in the early novels discovering who he is, Hester Latterly has no such identity crisis. Not only is she sure of who she is, her character often flies in the face of the socially accepted roles for Victorian Englishwomen: "She was highly intelligent, with a gift for logical thought which many people found disturbing—especially men, who did not expect it or like it in a woman" (*The Face of a Stranger*, p. 138).

The use of Hester as an icon for the changing role of women in English society is set against a background of an archaic class system that is also undergoing radical changes following the Crimean War. As a portrait of mid-nineteenth-century Victorian England emerges in the series, one recurring theme is the impact of war (in this case, the Crimean War) on English society. These commentaries could be applied just as easily to most wars. For many, substituting Vietnam and America for Crimea and England would ring disturbingly true.

> If he had ever known the purpose of war in the Crimea he had forgotten it now. It could hardly have been a war of defense. Crimea was a thousand miles from England. Presumably from the newspapers it was something to do with the political ramifications of Turkey and its disintegrating empire. It hardly seemed a reason for the wretched, pitiful deaths of so many, and the grief they left behind. (*The Face of a Stranger*, pp. 201–202)

Because Hester Latterly had served as a nurse in the Crimea with Florence Nightingale and had seen the senselessness and futility of the war, and because she is such a different kind of Englishwoman, intelligent and outspoken, Perry uses her often to comment on the idiocy of war and the paltry conditions of medical care in England, always suggesting that much of the problems result from a male-dominated establishment. She is highly critical of the commanders in the Crimea, whose stupidity and pride resulted in unnecessary suffering and death. As for the state of medical care in England, male doctors controlled the system, nurses were merely there for menial tasks, and the medical profession was highly resistant to change: "The medical establishment was desperately conservative, jealous of its knowledge and privilege, loathing change. There was no place for women except as drudges" (*A Sudden, Fearful Death*, p. 64).

Our introduction to medical care in mid-nineteenth-century England begins with Monk's observations in the hospital upon gaining consciousness from his accident.

> In the short days he could remember he had seen doctors move from one bloody or festering wound to another, from fever pa-

tients to vomiting and flux, then to open sores, and back again. Soiled bandages lay on the floor; there was little laundry done. (*The Face of a Stranger,* p. 8)

To emphasize the resistance of doctors to change or to the increasing role of nurses, Hester loses her first job upon returning from the Crimea when she saves a patient's life by applying procedures against the orders of the doctor. Even though the patient would have died, her insubordination cannot be tolerated.

Yet there have been recent advances in medical technology, and from time to time, they are interjected into the descriptions: "'Just think how hopeless that would have been twelve years ago before anesthetic. Now with ether or nitrous oxide, nothing is impossible'" (*A Sudden, Fearful Death,* p. 68).

Numerous other innovations are slipped in to remind us that Victorian England, for all its conservatism and resistance to change, was undergoing radical changes in all areas of life. This, then, provides the antagonism for the series—change versus resistance to change—as Anne Perry paints a diverse and dynamic portrait of life in England in the mid-nineteenth century as a backdrop for Victorian murders. ■■■■■■■■

The Turn-of-the-Century Cairo of Michael Pearce

[I]mmediate policing was the preserve of the police.
The Mamur Zapt, however, was responsible
for the preservation of order in the city
and Owen took the view that the first thing was
to get on top of any disorder and argue
about the division of responsibility afterwards.
(The Camel of Destruction, p. 92)

ONCE AGAIN WE NEED to stretch the criteria to justify the inclusion of the Mamur Zapt series as a police procedural. Captain Owen (his first name seems to change over time) is the Mamur Zapt, a position "roughly the equivalent of the Political Branch of the CID

[Criminal Investigations Division] in England." He is interested in the political ramifications of criminal activity in British-controlled Egypt and might be described as the head of Cairo's secret police. The investigation of crimes for the purpose of apprehending criminals is the responsibility of the Department of Prosecutions, known as the *Parquet,* in the Ministry of Justice, although the lines of responsibility are not always clearly delineated.

Edwardian Egypt was administered by a dual system of control, characterized by Egyptian ministers and ministries, all of which had their English "advisors." Even the Egyptian prime minister felt the watchful gaze of his British overlords.

> The Prime Minister . . . found it politic to draw abundantly on the wisdom of the [British] Consul-General before adopting a course of action. The system worked surprisingly well. From the British point of view, of course. (*The Camel of Destruction,* p. 15)

The basic nature of the relationship between Britain and its colony in northern Africa sets the stage for one of the dominant subplots of the series, the antagonisms between rulers and ruled. The plot for *The Mamur Zapt and the Donkey-Vous* revolves around the theme of striking out at the colonial oppressors, both British and French. Abductions occur out in the open on the terrace of the famous Shepheard's Hotel, a favorite spot for foreigners in Cairo. Why? The answer eventually comes to the Mamur Zapt.

> That was it! Why Moulin? Because he was French. Why Colthrope Hartley? Because he was British. Why Shepheard's? Because everyone could see.
>
> You could strike back at the oppressors. That was the lesson . . . Shepheard's! The very symbol of foreign privilege! The terrace! The most conspicuous place in Cairo. (*The Mamur Zapt and the Donkey-Vous,* p. 165)

Although Egypt was under British rule by this time, many features of French culture were present in Egyptian society, not the least of which was a judicial system based on the French Code Napoleon. Primarily an indigenous organization,

> as in France, the Parquet had the responsibility not just of investigating but also of preparing the case and carrying through any

> prosecution. The court often decided issues on the basis of the Parquet's report, or *procès-verbal,* rather than on the basis of testimony in court, which in Egypt was probably wise. (*The Mamur Zapt and the Men Behind,* pp. 37–38)

But the imprint of French culture went beyond the legacy of legal institutions. It also served as a sign of elite status within Egyptian society: "Like most of the Egyptian upper class, the Pasha habitually spoke French. He looked on the French culture as his own" (*The Mamur Zapt and the Men Behind,* p. 13).

Because there are no series based on officers of the turn-of-the-century Egyptian *Parquet,* the Mamur Zapt series is the only one that offers readers a place and time not usually employed as a setting for murder mysteries and worthy of consideration. Even though the Mamur Zapt is an agent of the dominant colonial power, he offers a perspective that is respectful of and sympathetic to indigenous peoples and cultures. He also offers multiple images of a dynamic world city. Cairo in the first decade of the twentieth century was "a city with over twenty different nationalities, at least five major religions apart from Islam, three principal languages, four competing legal systems and, in effect, two Governments" (*The Mamur Zapt and the Men Behind,* p. 19)—a diverse and cosmopolitan city in the latter stages of British colonialism.

The series is set in a tumultuous political environment, which captures the essential features of colonialism in its dying days. Pearce combines the intrigue and confusion that resulted from years of struggle between traditional indigenous, usually authoritarian rulers, British overlords during the period of colonialism, and the recently emerging forces of nationalism who most often were intent upon overthrowing the traditional system of rule, expelling the European colonial power, and establishing some kind of free, democratic society in its place. In most cases, the nationalists focused public dissatisfaction and unrest on the collaboration between the traditional rulers and the colonial power—such is the case with British Cairo in the Pearce series.

The stage is set with an overview of the changing nature of British rule in the early twentieth century. Pearce uses a change of consuls-general to underscore a liberalization of British policy toward Egypt, a liberalization that promoted the greater inclusion of Egyptians in the governance and administration of the colony. The recently replaced consul-general ruled Egypt with an iron fist for nearly thirty years, while the newly appointed consul-general epitomized the philosophy of the liberal parliament in England, which had recently replaced the long-standing conservative parliament.

At the same time, a crisis in the parallel indigenous government, ruled by a khedive, the hereditary ruler of Egypt, and his prime minister, forms the basis for court intrigue and the setting for at least one mystery in *The Mamur Zapt and the Men Behind.* The crisis results from the indecision of the khedive to appoint a new prime minister. The antagonisms that arise because the previous prime minister was Coptic Christian in a predominantly Moslem society raises other issues of ethnic diversity and conflict in this region of the world. The collaboration between the British and the khedive serves as a constant reminder of the way in which the colonial powers had promulgated the authoritarian status of the indigenous elite during colonialism. ▪▪▪▪▪▪▪▪▪▪▪▪▪

Conclusion

Taking murder mysteries back in time is one of the more difficult challenges facing a writer of fiction. In the four series examined here, the challenge has been met, and these authors have produced series that are fascinating not only because of their imaginative plot formulations, but also because of the seemingly authentic places used for these plots. In all cases, it is apparent that a great deal of research has gone into uncovering what these societies were like during different historical periods. Although there is no way to verify the correlation between the real and the fictional, these series give the impression that this is how it could have been and possibly was. We are left with some carefully crafted examples of the fictional union between history and geography.

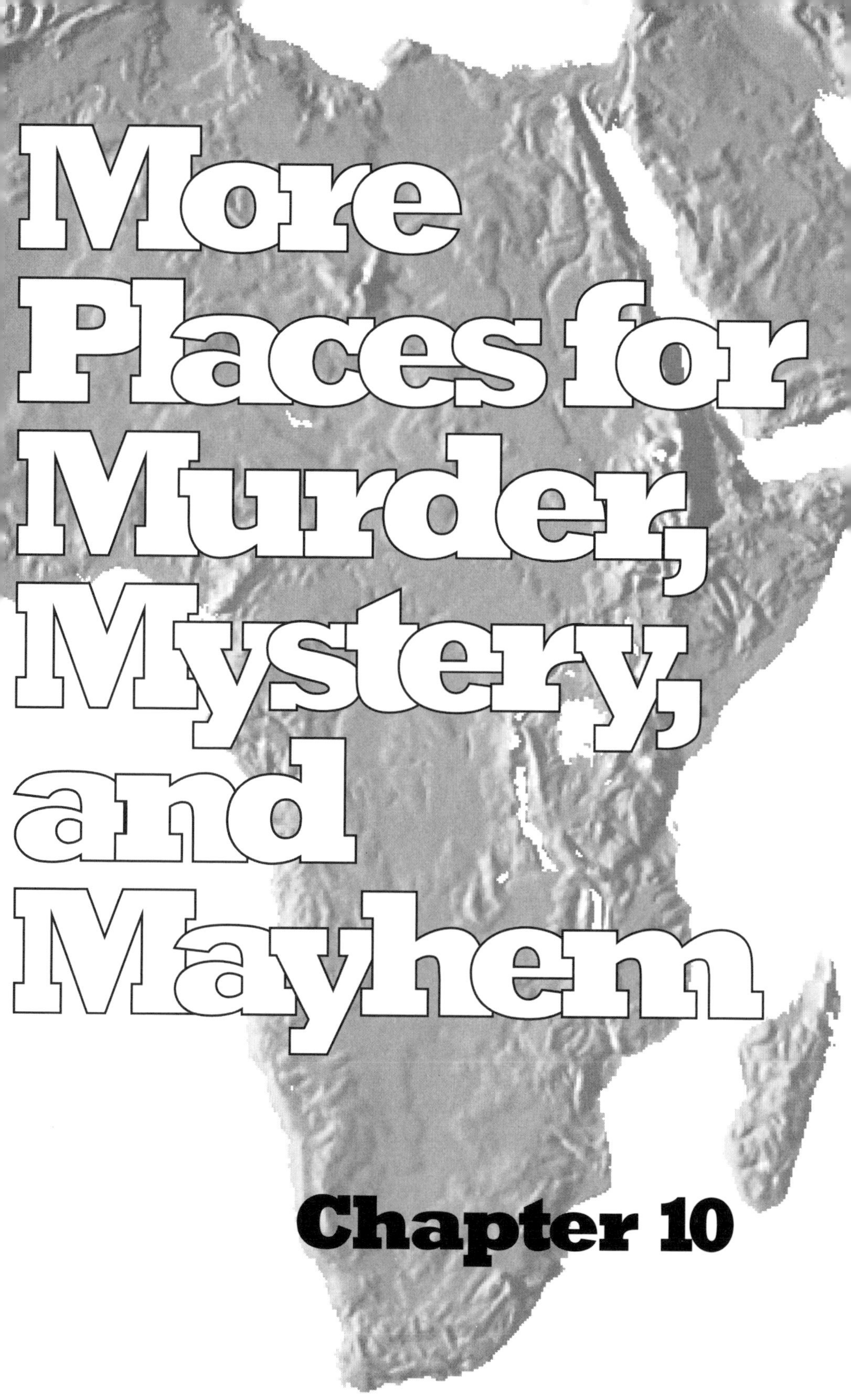

More Places for Murder, Mystery, and Mayhem

Chapter 10

In that well-known and controlled landscape
of the imagination the tensions, ambiguities,
and frustrations of ordinary experience are painted
over by magic pigments of adventure,
romance and mystery. The world for a time
takes on the shape of our heart's desire.
—JOHN CAWELTI, Adventure, Mystery, and Romance[1]

If nothing else, the examples used for this study underscore the great diversity, unbridled imaginativeness, and evident literary skills that epitomize the police procedural genre. This excursion not only further expands our understanding of place, it exposes the power of popular, escapist literature to influence, if not form, our images of places. It reinforces the premise that many police procedurals are fundamentally place-based. This survey also raises a number of provocative questions. Why is the police procedural so effective in conveying a sense of place? How do subplots and secondary agenda fit into the mix of place-based novels? What is the effect when authors get it wrong? How does "sense of place" in police procedurals fit into the overall context of socially contingent places; in other words, whose places are these?

THE POLICE PROCEDURAL AS AN EFFECTIVE CONVEYOR OF PLACE

One of the basic premises of this study is that the police procedural format is particularly conducive to effectively convey a sense of place, and that when this is done, place becomes an essential plot element. The thirty-three series selected for this examination reveal that being place-based is part of the overall structure for creating reality-based settings; this was the primary stimulus for the evolution of this genre of the murder mystery—that is, the demand for mysteries that were more lifelike in nature. Various literary devices, including narrative description, dialog, iconography, and attention to detail, work together to facilitate the creation of real-life places for murder. In some series, one or more of these devices proved stronger than in others; in some, an intermingling of all four are used to further the sense of place.

One consequence of these devices is an appreciation for the multifaceted nature of sense of place—it is seen as more than simply a description of the landscape. Place includes all of those physical and cultural attributes that distinguish it from all other places. This is the reason why place emerges as an essential plot element in placed-based novels.

One could make a similar case for these novels being character-based, procedural-based, socially based, and temporally based as well. To a certain extent, a place-based approach includes these other perspectives. All of these components work together to accommodate the generation of reality-based literary settings that impress readers with their authenticity and vibrancy.

SUBPLOTS AND SECONDARY AGENDA

Although numerous subplots and secondary agenda can be identified readily in these series, one common feature is that the subplots do not overpower the primary focus of the novel—the resolution of a murder. Different series show sensitivity to environmental issues, the role of women in society, ethnocentrism and racism, colonialism and post-colonialism, the plight of the disadvantaged, alcoholism and other kinds of abuse, and the disadvantages of incumbent political systems, to name the more prominent. In keeping with a place-based approach, it is worth noting that, to a great extent, the subplots of these novels tend to be regionally specific. The environment appears prominently in series based in the American West, racism in multicultural societies, especially Native American, African, Asian, and Australian, the role of females in series with female police officers, and colonialism and post-colonialism in series based in countries once part of the British Empire, or currently under English rule. In many cases, these represent more than just subplots; they are intrinsic components of the place and, thus, essential ingredients in understanding these places.

WHAT HAPPENS WHEN AUTHORS GET THEIR PLACES WRONG?

One of the difficulties that appears in several of these cultural murder mysteries are inconsistencies and errors about the places. Do these errors make the sense of place conveyed in these novels unreliable? Such criticism can be made about any genre of literature, no matter how serious the mode of conveyance. A sense of place is always merely the author's interpretation of the place, subject to all kinds of biases and idiosyncrasies. Although the reader needs to be aware of this, it does not necessarily diminish the usefulness of the interpretation. The point should be, does the series capture the essence of the place?

Even as renowned an author as Tony Hillerman has been cited for inaccuracies and inconsistencies in his depiction of Southwestern Native American culture. Although his friend and critic Ernie Bulow is able to point out numerous of these errors, he is quick to advise us not to be overly concerned about the minor ones Hillerman makes when describing Navajo culture, for example. The effectiveness of the novels results from the fact that Hillerman has "an unerring talent for invoking the mood, using a small bit of culture to weave a convincing tapestry of complex design and intricate pattern . . . Readers usually finish one of his books thinking they have been deeply immersed in Navajo culture."[2] The key to the place-based police procedural is conveying the essence of the place.

If the test lies in creating the essence of Soviet society and life in Moscow, for another example, the novels of Smith and Kaminsky pass with flying colors. Certainly, there are shortcomings; however, these do not detract substantially from the overall effectiveness. Kaminsky's work, for example, contains a few glitches. In addition to placing obvious St. Petersburg landmarks in Moscow, as already commented upon in Chapter 6, he has Sasha Tkach taking the Moscow metro at 3 a.m., after explaining to us that the metro closes at 1 a.m., which is true. Unfortunately, it does not reopen until 6 a.m.—a minor point, but an inaccuracy nonetheless. Also, those familiar with Russian names may be put off by his giving a male victim a female last name—Savitskaya. It should have been Savitsky. But all of these are rather minor glitches; they do not necessarily detract from the effectiveness of these novels. The more disturbing error, as discussed in Chapter 6, concerns placing street scenes from St. Petersburg in Moscow. As serious as this may be, it does not detract from the plot nor interfere with our ability to get a sense of Moscow.

Even Susan Dunlap, who lives in the Bay Area, refers to the Sierra as the "Sierras," an absolutely unforgivable sin to locals. Grating as that might be, such an error can be overlooked, if not entirely forgiven, because the essence of Berkeley comes through quite strikingly, and this faux pas does not detract from the overall sense of Berkeley or from the plot of the novel.

SOCIALLY CONTINGENT PLACES

The question of reliability raises another equally important question—whose places are being depicted? The basic tenets of the police procedural genre dictate that all of these series have certain commonalities, as originally discussed in Chapter 2; yet we have just been exposed to the diversity within the genre. Even for series based on similar character types, same time periods, and the same geographical regions, very different kinds of novels are being written. For our particular concern with the conveyance of place, there are two major sources for the different perspectives present in fiction: the author and the characters.

One of the challenges faced by authors is to imagine different perspectives and then present them through the iconography of the characters. The cop and the criminal, for example, will have very different interpretations of the same place, as will characters of different race, ethnicity, and gender, or different social status, wealth, and power. Even characters who appear so similar as Joe Leaphorn and Jim Chee are different kinds of Navajo, each of whom provides a different interpretation of Navajo culture and life on the reservation. As a result, they also bring to the plot different approaches to police work. In the person of "Bony" Bonaparte, we find two different interpretations through the eyes of the same person. In all of the series, characters as icons allow authors to voice these different perceptions of the same places. Yet this is the author's interpretation of each icon's perception of place, which brings us to the major source of differences in interpretation and perception—the author.

Clearly the most influential factor in determining the perspective taken by a particular piece of fiction derives from the author. For our concern with place, familiarity with the places used as settings for murder is of key interest, and the series examined here include authors with varied degrees of familiarity with their places. At one end of the spectrum, we have authors who have little first-hand knowledge of their places, authors like Martin Cruz Smith, Stuart Kaminsky, and H. R. F. Keating. Our four historical authors are familiar with their places but at drastically different historical periods. On the other hand, many of these authors write about places where they have lived or spent a great deal of time, authors like Tony Hillerman, Susan Dunlap, James Lee Burke, and Peter Turnbull.

One impression comes through in almost all cases. These authors are writing about places for which they have a strong attachment. Often presented in terms of a love-hate relationship, where the different faces of the locale are presented, authors do not tend to write, or so it would seem, about places they do not like. As a result, we are given rather sympathetic introductions to exotic places through the interpretations of authors who do not just know their places, do not just feel comfortable writing about their places, but have a genuine affection for those places.

THE POLICE PROCEDURAL AS A SOURCE OF SENSE OF PLACE

In the end, we have had more than just a good read and a solved murder. We have gained a sense of place that has been fed to us throughout the novel. Sometimes the appreciation of city and country was fed to us unnoticed; other times it was inescapable for plot development. A valuable source of literary geography and a widely accepted sense of place is contained within the pages of these popular novels. It still appears that academics, particularly geographers, have continued to ignore this genre of literature because it is not "serious" literature. This is not as true for the humanities where numerous scholars have appreciated the necessity of dealing with popular novels seriously, whether or not they are serious literature. This is all the more alarming because of the transcendent nature of literature. Sense of place in literature, especially in the cultural murder mystery novel, binds together the humanities, the social sciences, and even the natural sciences. Geographers have a great deal to contribute to the discussion, especially in taking the discussion of sense of place beyond the simple understanding of the importance of setting and mood. Who better to address the fuller ramifications and subtleties of place than the geographer?

And where better to understand contemporary society than through the eyes of good novelists writing mystery novels, novels that are rooted in the very fabric of society? The very nature of places is embodied in the vibrant and popular murder mystery novel, where the impact is so indelibly etched in the minds of the readers because it is a mystery novel first, an adventure into sense of place second.

In fact, the very nature of this genre, and the relatively large audience it reaches, guarantees that it is probably more effective than many kinds of "serious" literature. Like all murder mysteries, the place-based police procedural presents a relatively underutilized source for literary geography and the study of sense of place in literature. Certainly there should be floor room enough in the house of geography for more dead bodies. The murder mystery has never been an underutilized source for public awareness of sense of place. Now, the place-based police procedural increases the power of the genre to convey this most essential geographic theme.

Appendix: Selected Series

AMERICA

Burke, James Lee (New Orleans: Dave Robicheaux)

The Neon Rain, 1987.
Heaven's Prisoners, 1988.
Black Cherry Blues, 1989.
A Morning for Flamingos, 1990.
A Stained White Radiance, 1992.
In the Electric Mist with Confederate Dead, 1993.
Dixie City Jam, 1994.
Burning Angel, 1995.
Sunset Limited, 1998.

Carlson, P. M. (Indiana: Marty Hopkins)

Gravestone, 1993.
Bloodstream, 1995.

Dunlap, Susan (Berkeley: Jill Smith)

Karma, 1981.
As a Favor, 1984.
Not Exactly a Brahmin, 1985.
Too Close to the Edge, 1987.
A Dinner to Die For, 1987.
Diamond in the Buff, 1990.
Death and Taxes, 1992.
Time Expired, 1993.
Sudden Exposure, 1996.
Cop Out, 1997.

Hager, Jean (Cherokee country: Mitchell Bushyhead)

The Grandfather Medicine, 1990.
Ghostland, 1991.

Night Walker, 1991.
The Fire Carrier, 1996.
Masked Dancers, 1998.

Hillerman, Tony (Navajo country: Jim Chee and Joe Leaphorn)

The Blessing Way, 1970.
Dance Hall of the Dead, 1973.
Listening Woman, 1978.
People of Darkness, 1980.
The Dark Wind, 1982.
The Ghostway, 1985.
Skinwalkers, 1987.
A Thief of Time, 1988.
Talking God, 1989.
Coyote Waits, 1990.
Sacred Clowns, 1993.
The Fallen Man, 1996.
The First Eagle, 1998.

Jance, J. A. (Seattle: J. P. Beaumont)

Until Proven Guilty, 1985.
Injustice for All, 1986.
Taking the Fifth, 1987.
Dismissed With Prejudice, 1989.
A More Perfect Union, 1990.
Payment in Kind, 1991.
Without Due Process, 1992.
Lying in Wait, 1992.
Failure to Appear, 1993.
Name Withheld, 1996.
Minor in Possession, 1997.
Trial by Fury, 1997.
Improbable Cause, 1997.
Breach of Duty, 1999.

Lindsey, David (Houston: Stuart Haydon)

Heat From Another Sun, 1984.
Spiral, 1986.
In the Lake of the Moon, 1987.
Body of Truth, 1992.
A Cold Mind, 1993.

Smith, Julie (New Orleans: Skip Langdon)

New Orleans Mourning, 1990.
The Axeman's Jazz, 1991.
Jazz Funeral, 1993.
New Orleans Beat, 1994.
House of Blues, 1995.
Kindness of Strangers, 1996.
Crescent City Kill, 1997.
82 Desire, 1998.

Taibo II, Paco Ignacio (Mexico)

Life Itself. Translated by Beth Hanson, 1990.
Some Clouds. Translated by William Neuman, 1992.
No Happy Ending. Translated by William Neuman, 1993.
Leonardo's Bicycle. Translated by Martin Roberts, 1993.
Four Hands. Translated by Laura Dial, 1994.

Young, Scott (Canada: Matteesie Kitologitak)

Murder in a Cold Climate, 1989.
The Shaman's Knife, 1993.

UNITED KINGDOM AND IRELAND

Dexter, Colin (Oxford: Inspector Morse)

Last Bus to Woodstock, 1975.
Last Seen Wearing, 1976.
The Silent World of Nicholas Quinn, 1977.
Service of All the Dead, 1979.
The Dead of Jericho, 1981.
The Riddle of the Third Mile, 1983.
The Secret of Annexe 3, 1986.
The Wench is Dead, 1989.
The Jewel That Was Ours, 1991.
The Way Through the Woods, 1992.
The Daughters of Cain, 1994.
Death is Now My Neighbor, 1996.

Gill, Bartholomew (Ireland: Peter McGarr)

McGarr and the Politician's Wife, 1977.
McGarr and the Sienese Conspiracy, 1977.
McGarr on the Cliffs of Moher, 1978.
McGarr at the Dublin Horse Show, 1980.

McGarr and the P. M. of Belgrave Square, 1983.
McGarr and the Method of Descartes, 1985.
McGarr and the Legacy of a Woman Scorned, 1986.
The Death of a Joyce Scholar, 1989.
The Death of Love, 1992.
Death on a Cold, Wild River, 1993.
The Death of an Ardent Bibliophile, 1995.
The Death of an Irish Seawolf, 1996.
The Death of an Irish Tinker, 1997.

James, P. D. (London: Adam Dalgliesh)

Cover Her Face, 1962.
A Mind to Murder, 1963.
Unnatural Causes, 1967.
Shroud for a Nightingale, 1971.
The Black Tower, 1975.
Death of an Expert Witness, 1977.
A Taste for Death, 1986.
Devices and Desires, 1990.
Original Sin, 1994.
A Certain Justice, 1997.

Robinson, Peter (Yorkshire: Alan Banks)

Gallow's View. 1987.
A Necessary End, 1989.
The Hanging Valley, 1994.
Past Reason Hated, 1994.
Final Account, 1994.
Wednesday's Child, 1995.
Innocent Graves, 1996.
Blood at the Root, 1997.
In a Dry Season, 1999.

Turnbull, Peter (Scotland: P Division)

Deep and Crisp and Even, 1981.
Dead Knock, 1982.
Fair Friday, 1983.
Big Money, 1984.
Two Way Cut, 1988.
Condition Purple, 1989.
Long Day Monday, 1992.
The Killing Floor, 1994.
The Man with No Face, 1998.

EUROPEAN MAINLAND

Dibdin, Michael (Italy: Aurelio Zen)

Ratking, 1989.
Vendetta, 1991.
Cabal, 1994.
Dead Lagoon, 1994.
Cosi Fan Tutti, 1996.
A Long Finish, 1998.

Freeling, Nicolas (Northern France: Henri Castang)

A Dressing of Diamond, 1974.
The Bugles Blowing, 1976.
Sabine, 1978.
The Night Lords, 1978.
Castang's City, 1980.
Wolfnight, 1982.
The Back of the North Wind, 1983.
No Part in Your Death, 1984.
A City Solitary, 1985.
Cold Iron, 1986.
Lady Macbeth, 1988.
Not as Far as Velma, 1989.
Flanders Sky, 1992.
The Seacoast of Bohemia, 1994.
You Who Know, 1994.
A Dwarf Kingdom, 1996.

van de Wetering, Janwillem (Amsterdam: Grijpstra and de Gier)

Outsider in Amsterdam, 1975.
Tumbleweed, 1976.
The Corpse on the Dike, 1976.
Death of a Hawker, 1977.
The Japanese Corpse, 1977.
The Blond Baboon, 1978.
The Main Massacre, 1979.
The Mind-Murders, 1981.
The Streetbird, 1983.
The Rattle-Rat, 1985.
Hard Rain, 1986.
Just a Corpse at Twilight, 1994.
The Perfidious Parrot, 1998.

Wahloo, Per and Sjowall, Maj (Stockholm: Martin Beck)

Roseanna, 1967.
The Man on the Balcony, 1968.
The Man Who Went Up in Smoke, 1969.
The Laughing Policeman, 1970.
The Fire Engine That Disappeared, 1971.
Murder at the Savoy, 1971.
The Abominable Man, 1972.
The Locked Room, 1973.
Cop Killer, 1975.
The Terrorists, 1976.

MOSCOW

Kaminsky, Stuart (Moscow: Porfiry Rostnikov)

Death of a Dissident, 1981.
Black Knight in Red Square, 1984.
Red Chameleon, 1985.
A Fine Red Rain, 1987.
A Cold Red Sunrise, 1988.
The Man Who Walked Like a Bear, 1990.
Rostnikov's Vacation, 1991.
Death of a Russian Priest, 1992.
Hard Currency, 1994.
Blood and Rubles, 1995.
Tarnished Icons, 1997.

Smith, Martin Cruz (Moscow: Arkady Renko)

Gorky Park, 1981.
Polar Star, 1989.
Red Square, 1992.
Havana Bay, 1999.

THE ORIENT

Marshall, William (Hong Kong: Harry Feiffer)

Yellowthread Street, 1976.
Gelignite, 1977.
The Hatchet Man, 1977.
Thin Air, 1978.
Skulduggery, 1980.
Sci Fi, 1981.
Perfect End, 1983.

War Machine, 1988.
The Faraway Man, 1985.
Road Show, 1985.
Head First, 1986.
Frogmouth, 1987.
Out of Nowhere, 1988.
Inches, 1994.
Nightmare Syndrome, 1997.
To the End, 1998.

Matsumoto, Seicho (Japan)

Inspector Imanishi Investigates, 1961.
Points and Lines, 1970.

Melville, James (Japan: Tetsuo Otani)

Wages of Zen, 1979.
The Chrysanthemum Chain, 1980.
A Sort of Samurai, 1981.
The Ninth Netsuke, 1982.
Sayonara, Sweet Amaryllis, 1983.
Death of a Daimyo, 1984.
The Death Ceremony, 1985.
Go Gently, Gaijin, 1986.
Kimono for a Corpse, 1987.
The Reluctant Ronin, 1988.
The Bogus Buddha, 1990.
The Body Wore Brocade, 1992.

West, Christopher (Beijing: Wang Anzhuang)

Death of a Blue Lantern, 1994.
Death on Black Dragon River, 1999.

OTHER PLACES

Gur, Batya (Israel: Michael Ohayon)

Saturday Morning Murder, 1992.
Literary Murder, 1993.
Murder on a Kibbutz, 1994.
Murder Duet, 1999.

Keating, H. R. F. (Bombay: Ganesh Ghote)

The Perfect Murder, 1964.
Inspector Ghote's Good Crusade, 1966.

Inspector Ghote Caught in Meshes, 1967.
Inspector Ghote Hunts the Peacock, 1968.
Inspector Ghote Plays a Joker, 1969.
Inspector Ghote Breaks an Egg, 1971.
Inspector Ghote Goes by Train, 1972.
Inspector Ghote Trusts the Heart, 1973.
Bats Fly Up for Inspector Ghote, 1974.
Filmi, Filmi, Inspector Ghote 1977.
Inspector Ghote Draws a Line, 1979.
Go West, Inspector Ghote, 1981.
The Sheriff of Bombay, 1984.
Under a Monsoon Cloud, 1986.
Dead on Time, 1989.
Inspector Ghote, His Life and Crimes, 1989.
The Iciest Sin, 1990.
Cheating Death, 1992.
Doing Wrong, 1994.
Asking Questions, 1996.

McClure, James (South Africa: Tromp Kramer and Mickey Zondi)

The Steam Pig, 1972.
The Caterpillar Cop, 1973.
The Gooseberry Fool, 1974.
Snake, 1976.
The Sunday Hangman, 1977.
The Blood of an Englishman, 1981.
The Artful Egg, 1985.
The Song Dog, 1991.

Upfield, Arthur (Australia: Napoleon "Bony" Bonaparte)

The Lure of the Bush, 1929.
The Sands of Windee, 1931.
Wings Above the Claypan, 1943.
Murder Down Under, 1943.
The Mystery of Swordfish Reef, 1943.
Wind of Evil, 1944.
No Footprints in the Bush, 1944.
Death of a Swagman, 1945.
The Devil's Steps, 1946.
The Bone is Pointed, 1947.
An Author Bites the Dust, 1948.
The Mountains Have a Secret, 1948.
The Widows of Broome, 1950.

The Bachelors of Broken Hill, 1950.
The New Shoe, 1951.
Venom House, 1952.
Murder Must Wait, 1953.
Death of a Lake, 1954.
Sinister Stones, 1954.
The Battling Prophet, 1956.
The Man of Two Tribes, 1956.
The Bushman Who Came Back, 1957.
Journey to the Hangman, 1959.
Valley of Smugglers, 1960.
The White Savage, 1961.
The Will of the Tribe, 1962.
The Body at Madman's Bend, 1963.
Bony and the Black Virgin, 1965.
The Lake Frome Monster [completed by Price and Strange], 1966.

HISTORIC PLACES

Davis, Lindsey (First-century Rome: Marcus Didius Falco)

Silver Pigs, 1991.
Venus in Copper, 1991.
Shadows in Bronze, 1992.
The Iron Hand of Mars, 1992.
Poseidon's Gold, 1992.
Last Act in Palmyra, 1994.
Time to Depart, 1995.
A Dying Light in Corduba, 1998.
Three Hands in the Fountain, 1999.

Pearce, Michael (Turn-of-the-century Cairo: the Mamur Zapt)

The Mamur Zapt and the Return of the Carpet, 1990.
The Mamur Zapt and the Night of the Dog, 1991.
The Mamur Zapt and the Men Behind, 1991.
The Mamur Zapt and the Donkey-Vous, 1992.
The Camel of Destruction, 1993.
The Mamur Zapt and the Girl in the Nile, 1994.
The Mamur Zapt and the Spoils of Egypt, 1995.
The Snake-Catcher's Daughter, 1998.

Perry, Anne (Victorian England: William Monk)

The Face of a Stranger, 1990.
A Dangerous Mourning, 1991.

Defend and Betray, 1992.
A Sudden, Fearful Death, 1993.

Van Gulik, Robert (Seventh-century China: Judge Dee)

Celebrated Cases of Judge Dee (Dee Goong An). Translated and with an introduction and notes by Robert Van Gulik, 1949.
The Chinese Gold Murders, 1959.
The Chinese Bell Murders, 1960.
The Chinese Lake Murders, 1960.
The Haunted Monastery, 1961.
The Red Pavilion, 1961.
The Chinese Nail Murders, 1961.
The Chinese Maze Murders, 1962.
The Lacquer Screen, 1962.
The Emporer's Pearls, 1963.
The Willow Pattern, 1965.
The Monkey and the Tiger, 1965.
Murder in Canton, 1966.
The Phantom of the Temple, 1966.
Judge Dee at Work, 1967.
Necklace and Calabash, 1967.
Poets and Murder, 1968.

Fictional Works Cited

Publishing information for nonfiction sources cited in this work can be found in the Notes section.

AMERICA

Burke, James Lee

Burning Angel. Hyperion, 1995.
Dixie City Jam. Hyperion, 1994.
Heaven's Prisoners. Pocket Books, 1989.
The Neon Rain. Pocket Books, 1988.

Carlson, P. M.

Bloodstream. Pocket Books, 1996.
Gravestone. Pocket Books, 1994.

Constantine, K. C.

Bottom Line Blues, The Mysterious Press, 1994.
Sunshine Enemies. The Mysterious Press, 1991.

Dunlap, Susan

A Dinner to Die For. New York: Dell, 1990.
Not Exactly a Brahmin. New York: Dell, 1991.
Sudden Exposure. New York: Dell, 1997.

Hager, Jean

The Fire Carrier. The Mysterious Press, 1997.
The Grandfather Medicine. Worldwide, 1990.

Hillerman, Tony

Coyote Waits. Harper, 1992.
Dance Hall of the Dead. Harper, 1990.
Skinwalkers. Harper, 1990.

Talking God. Harper, 1991.
A Thief of Time. Harper, 1990.

Jance, J. A.

Failure to Appear. Avon, 1994.
Injustice for All. Avon, 1986.
Taking the Fifth. Avon, 1987.

Lindsey, David

Body of Truth. Bantam, 1993.
A Cold Mind. Pocket Books, 1984.
In the Lake of the Moon. Bantam, 1990.

McBain, Ed

Romance. Warner, 1995.

Smith, Julie

The Axeman's Jazz. Ivy, 1992.
Jazz Funeral. Ivy, 1994.
Kindness of Strangers. Ivy, 1997.

Taibo II, Paco Ignacio

Life Itself. Translated by Beth Henson. The Mysterious Press, 1995.

Waugh, Hillary

Last Seen Wearing. Doubleday, 1952.

Young, Scott

The Shaman's Knife. Penguin, 1994.

UNITED KINGDOM AND IRELAND

Dexter, Colin

The Jewel That Was Ours. Ivy, 1993.
Last Bus to Woodstock. Ivy, 1996.
Last Seen Wearing. Bantam, 1989.
The Secret of Annexe 3. Bantam, 1988.
The Silent World of Nicholas Quinn. Ivy, 1997.
The Wench is Dead. Bantam, 1991.

Gill, Bartholomew

The Death of an Ardent Bibliophile. Avon, 1996.
The Death of an Irish Tinker. Avon, 1998.
The Death of Love. Avon, 1993.

James, P. D.

Cover Her Face. Warner, 1982.
Devices and Desires. Warner, 1991.
A Mind to Murder. Warner, 1982.
Original Sin. Warner, 1996.

Robinson, Peter

Final Account. Berkeley Prime Crime, 1996.
The Hanging Valley. Berkeley Prime Crime, 1994.
A Necessary End. Penguin, 1990.
Past Reason Hated. Berkeley Prime Crime, 1994.
Wednesday's Child. Berkeley Prime Crime, 1995.

Turnbull, Peter

The Killing Floor. Worldwide, 1996.
Long Day Monday. Worldwide, 1995.

EUROPEAN MAINLAND

Dibdin, Michael

Cabal. Faber and Faber, 1992.
Dead Lagoon. Vintage Crime/Black Lizard, 1996.
Ratking. Vintage Crime/Black Lizard, 1997.

Freeling, Nicolas

The Bugles Blowing. Vintage, 1980.
Castang's City. Vintage, 1981.
A Dwarf Kingdom. The Mysterious Press, 1997.
Flanders Sky. The Mysterious Press, 1993.
Wolfnight. Vintage, 1983.

van de Wetering, Janwillem

The Corpse on the Dike. Soho, 1976.
Death of a Hawker. Pocket Books, 1978.
The Japanese Corpse. Soho, 1996.
Just a Corpse at Twilight. Soho, 1994.
The Maine Massacre. Ballantine, 1988.
Tumbleweed. Soho, 1992.

Wahloo, Per and Sjowall, Maj

The Laughing Policeman. Translated by Alan Blair. Vintage Crime/Black Lizard, 1992.
The Locked Room. Translated by Paul Britten Austin. Vintage Crime/Black Lizard, 1992.

The Man on the Balcony. Translated by Alan Blair. Vintage Crime/Black Lizard, 1993.
The Man Who Went Up in Smoke. Translated by Joan Tate. Vintage Crime/Black Lizard, 1993.
Murder at the Savoy. 1971.
Roseanna. Translated by Lois Roth. Vintage Crime/ Black Lizard, 1993.

MOSCOW

Kaminsky, Stuart

Blood and Rubles. Ivy, 1997.
A Cold Red Sunrise. Ivy, 1989.
Death of a Dissident. Ivy, 1989.
Death of a Russian Priest. Ivy, 1993.
A Fine Red Rain. Ivy, 1988.
Red Chameleon. Ivy, 1989.
Rostnikov's Vacation. Ivy, 1992.

Smith, Martin Cruz

Gorky Park. Ballantine, 1982.
Havana Bay. Random House, 1999.
Polar Star. Ballantine, 1989.
Red Square. Ballantine, 1993.

THE ORIENT

Marshall, William

The Faraway Man. Futura, 1987.
Inches. The Mysterious Press, 1994.
Nightmare Syndrome. The Mysterious Press, 1997.
Road Show. The Mysterious Press, 1988.
Thin Air. The Mysterious Press, 1978.
To the End. The Mysterious Press, 1998.

Matsumoto, Seicho

Inspector Imanishi Investigates. Translated by Beth Cary. Soho, 1989.
Points and Lines. Translated by Makiko Yamamoto and Paul Blum. Kodansha International, 1986.

Melville, James

The Body Wore Brocade. Scribner, 1992.
The Chrysanthemum Chain. Scribner, 1986.
Death of a Daimyo. Scribner, 1987.
The Wages of Zen. Scribner, 1985.

West, Christopher

Death of a Blue Lantern. Berkeley Prime Crime, 1998.

OTHER PLACES

Gur, Batya

Literary Murder. Translated by Dalya Bilu. Harper, 1994.
Murder Duet. Translator unknown at press time. HarperCollins, 1999.
Murder on a Kibbutz. Translated by Dalya Bilu. Harper, 1995.
The Saturday Morning Murder. Translated by Dalya Bilu. Harper, 1993.

Keating, H. R. F.

Asking Questions. St. Martin's, 1997.
Cheating Death. The Mysterious Press, 1994.
Doing Wrong. Otto Penzler Books, 1994.
The Iciest Sin. The Mysterious Press, 1991.
The Perfect Murder. Pan Books, 1996.

McClure, James

The Gooseberry Fool. Faber and Faber, 1993.
Snake. Faber and Faber, 1993.
The Song Dog. The Mysterious Press, 1992.
The Steam Pig. Faber and Faber, 1993.

Upfield, Arthur

Man of Two Tribes, 1956.
Murder Down Under. Scribner, 1983.
No Footprints in the Bush. Collier, 1986.
The Will of the Tribe, 1962.

HISTORIC PLACES

Davis, Lindsey

The Iron Hand of Mars. Ballantine, 1994.
Last Act in Palmyra. The Mysterious Press, 1997.
Silver Pigs. Ballantine, 1991.

Pearce, Michael

The Camel of Destruction. HarperCollins, 1995.
The Mamur Zapt and the Donkey-Vous. The Mysterious Press, 1993.
The Mamur Zapt and the Men Behind. The Mysterious Press, 1994.

Perry, Anne

The Face of a Stranger. Ivy, 1991.
A Sudden, Fearful Death. Ivy, 1994.

Van Gulik, Robert

Celebrated Cases of Judge Dee (Dee Goong An). Translated and with an introduction and notes by Robert Van Gulik. Dover Publications, 1976.
The Chinese Bell Murders. University of Chicago, 1977.
The Chinese Lake Murders. University of Chicago, 1979.
The Chinese Nail Murders. University of Chicago, 1977.
The Monkey and the Tiger. University of Chicago, 1992.
Murder in Canton. University of Chicago, 1993.
The Willow Pattern. University of Chicago, 1993.

Notes

NOTES TO CHAPTER 1

1. Jon L. Breen, "Introduction," in Ed Gorman, Martin Greenberg, and Larry Segriff, eds., with Jon L. Breen, *The Fine Art of Murder,* (New York: Carroll & Graf, 1993), p. 6.
2. C. Salter and W. Lloyd, *Landscape in Literature,* Resource Paper No. 76-3 (Washington, D.C.: AAG, 1977), p. 28.
3. For in-depth discussions of the escapist nature of murder mysteries, see: H. R. F. Keating, *Whodunit?: A Guide to Crime, Suspense and Spy Fiction* (New York: Van Nostrand Rheinhold, 1982), Mary Roth, *Foul and Fair Play: Reading Genre in Classic Detective Fiction* (Athens: University of Georgia Press, 1995), and John Ball, ed., *The Mystery Story* (San Diego: University of California Extension, 1976); for an introduction to the place-based procedural, see: Gary Hausladen, "Murder in Moscow," *Geographical Review* 85 (January 1995): 63–78, and "Where the Bodies Lie: Sense of Place and Police Procedurals," *Journal of Cultural Geography* 16 (Fall/Winter 1996): 45–63.
4. George Dove, "The Detective Formula and the Art of Reading," in Ronald Walker and June Frazier, eds., *The Cunning Craft: Original Essays on Detective Fiction and Contemporary Literary Theory* (Macomb, IL: Western Illinois University, 1990), p. 37.
5. Tony Hillerman, *Finding Moon,* (New York: Harper Collins, 1995), p. ix.
6. Phillip Plowden, "Marshall, William (Leonard)," in Lesley Henderson, ed., *Twentieth-Century Crime and Mystery Writers,* 3rd ed. (Chicago: St. Martin's Press, 1991), p. 719.

NOTES TO CHAPTER 2

1. Jon L. Breen, "Introduction" to the section "Police Procedural," in Gorman, Greenberg, and Segriff, eds., *The Fine Art of Murder,* p. 211.
2. Martin Priestman, *Detective Fiction and Literature: the figure on the carpet* (New York: St. Martin's Press, 1991), p. 17.

3. S. E. Sweeney, "Locked Rooms: Detective Fiction, Narrative Theory, and Self-reflexivity," in Walker and Frazer, eds., *The Cunning Craft,* p. 7.
4. J. R. Cox, *Masters of Mystery and Detective Fiction* (Pasadena: Salem Press, 1989), p. 1.
5. Fredric Jameson, "On Raymond Chandler," *Southern Review* 6 (1970), p. 626.
6. Hillary Waugh, "The Human Rather Than Superhuman Sleuth," in Lucy Freeman, ed., *The Murder Mystique: Crime Writers on Their Art* (New York: Frederick Ungar, 1982), p. 35.
7. George Dove, *The Police Procedural* (Bowling Green: Popular Press, 1982), p. 2.
8. Hillary Waugh, "The Police Procedural," in John Ball, ed., *The Mystery Story* (San Diego: University of California Extension, 1976), p. 167.
9. Jo Ann Vicarel, *A Reader's Guide to the Police Procedural* (New York: G. K. Hall & Co., 1995), p. x.
10. Hillary Waugh, "The American Police Procedural," in Keating, *Whodunit?,* p. 46.
11. Leroy Lad Panek, *An Introduction to the Detective Story* (Bowling Green: Popular Press, 1987), p. 187.
12. Robin Winks, *Modus Operandi: An Excursion into Detective Fiction* (Boston: David Godine, 1982), p. 61.
13. Winks, *Modus Operandi,* p. 92.
14. Julian Symons, *Bloody Murder: From the Detective Story to the Crime Novel* (New York: Mysterious Press, 1992), pp. 306–307.
15. James Shortridge, "The Concept of the Place-defining Novel in American Popular Culture," *Professional Geographer* 43 (1991), p. 280.
16. Shortridge, "The Place-defining Novel," p. 286.
17. John Agnew and James Duncan, "Introduction," in Agnew and Duncan, eds., *The Power of Place: Bringing Together Geographical and Sociological Imaginations* (Boston: Unwin Hyman, 1989), p. 2.
18. Douglas McManis, "Places for Mysteries," *Geographical Review* 68 (1978), p. 319.
19. Yi-Fu Tuan, "The Landscapes of Sherlock Holmes," *Journal of Geography* 84 (1985): 56–60.
20. Gorman, Greenberg, and Segriff, *The Fine Art of Murder,* p. 211.
21. Dennis Porter, *The Pursuit of Crime: Art and Ideology in Detective Fiction* (New Haven: Yale University Press, 1981), p. 187.
22. Porter, *The Pursuit of Crime,* p. 189.
23. Art Bourgeau, *The Mystery Lover's Companion* (New York: Crown, 1986), p. 277.
24. Winks, *Modus Operandi,* p. 58.
25. Seymour Chatman, *Coming to Terms: The Rhetoric of Narrative in Fiction and Film* (Ithaca: Cornell University Press, 1990).
26. Denis Cosgrove and Stephen Daniels, "Introduction: Iconography and Land-

scape," in Cosgrove and Daniels, eds., *The Iconography of Landscape* (Cambridge: Cambridge University Press, 1988), p. 2.

27. Priestman, *Detective Fiction and Literature,* p. ix.
28. Robin Woods, " 'His Appearance Is Against Him': The Emergence of the Detective," in Walker and Frazer, *The Cunning Craft,* p. 15.
29. Priestman, *Detective Fiction and Literature,* p. ix.
30. Jon Thompson, *Fiction, Crime, and Empire: Clues to Modernity and Postmodernism* (Urbana: University of Illinois Press, 1993), p. 8.
31. Ernest Mandell, *Delightful Murder: A Social History of the Crime Story* (Minneapolis: University of Minnesota Press, 1984), p. 9.
32. Mandell, *Delightful Murder,* p. 9.
33. Mandell, *Delightful Murder,* p. vi.
34. Robert Winston and Nancy Mallerski, *The Public Eye: Ideology and the Police Procedural* (New York: St. Martin's, 1992), p. 2.
35. E. T. Guyown, Jr., "Why Do We Read This Stuff?," in Keating, *Whodunit?,* p. 362.
36. Gary Hoppenstand, *In Search of the Tiger: A Sociological Perspective of Myth, Formula and the Mystery Genre in the Entertainment Print Mass Medium* (Bowling Green: Popular Press, 1987), p. 30.
37. John Cawelti, *Adventure, Mystery, and Romance: Formula Stories as Art and Popular Culture* (Chicago: University of Chicago Press, 1976), p. 77.
38. Cawelti, *Adventure, Mystery, and Romance,* p. 8.
39. Cawelti, *Adventure, Mystery, and Romance,* p. 43.
40. Porter, *The Pursuit of Crime,* p. 115.
41. Dove, *The Police Procedural,* p. 5.
42. Dove, *The Police Procedural,* p. 133.
43. For a more detailed account of how critics have attempted to explain detective fiction, see: Heta Pyrhonen, *Murder from an Academic Angle: An Introduction to the Study of the Detective Narrative* (Columbia, SC: Camden House, 1994).

NOTES TO CHAPTER 10

1. Cawelti, *Adventure, Mystery, and Romance,* p.1.
2. Tony Hillerman and Ernie Bulow, *Talking Mysteries: A Conversation with Tony Hillerman* (Albuquerque: University of New Mexico Press, 1991), p. 17.

Index